Engineering Fluid Mechanics

Engineering Fluid Mechanics

Editor
Azaire Ivanov

scitus
academics

Engineering Fluid Mechanics
Edited by **Azaire Ivanov**

ISBN: 978-1-68117-119-7
Library of Congress Control Number: 2015951142

SCITUS Academics LLC,
www.scitusacademics.com
Box No. 4766, 616 Corporate Way,
Suite 2, Valley Cottage,
NY 10989

Printed in United States of America on Acid Free Paper

Preface

Fluids are composed of molecules that collide with one another and solid objects. The continuum assumption, however, considers fluids to be continuous. Fluid mechanics is the branch of physics that studies the mechanics of fluids and the forces on them. Fluid mechanics can be divided into fluid statics, the study of fluids at rest; and fluid dynamics, the study of the effect of forces on fluid motion. Fluid mechanics, especially fluid dynamics, is an active field of research with many problems that are partly or wholly unsolved. Fluid mechanics can be mathematically complex, and can best be solved by numerical methods, typically using computers. A modern discipline, called computational fluid dynamics (CFD), is devoted to this approach to solving fluid mechanics problems. Particle image velocimetry, an experimental method for visualizing and analyzing fluid flow, also takes advantage of the highly visual nature of fluid flow. Fluid statics or hydrostatics is the branch of fluid mechanics that studies fluids at rest. It embraces the study of the conditions under which fluids are at rest in stable equilibrium; and is contrasted with fluid dynamics, the study of fluids in motion.

Hydrostatics is fundamental to hydraulics, the engineering of equipment for storing, transporting and using fluids. Fluid dynamics is a sub discipline of fluid mechanics that deals with fluid flow—the natural science of fluids (liquids and gases) in motion. Some of its principles are even used in traffic engineering, where traffic is treated as a continuous fluid, and crowd dynamics.

Fluid dynamics offers a systematic structure, which underlies these practical disciplines that embraces empirical and semi-empirical laws derived from flow measurement and used to solve practical problems. The solution to a fluid dynamics problem typically involves calculating various properties of the fluid, such as velocity, pressure, density, and temperature, as functions of space and time.

Fluid Mechanics is an essential subject in the study of the behaviour of fluids. The book is complimented by many worked examples, contains innovative ideas on fluid mechanics.

Table of Contents

CHAPTER 1

Numerical Investigation for Steady and Unsteady Cavitating Flows

Hatem Kanfoudi, Hedi Lamloumi and Ridha Zgolli

[1]Laboratory of Hydraulic and Environmental Modelling, National Engineering School of Tunis, Tunis, Tunisia

INTRODUCTION

Cavitation is a particular two-phase flow with phase transition (vaporization/condensation) driven by pressure change without any heating. It can be interpreted as the rupture of the liquid continuum due to excessive stresses. Modeling of cavitating flows as a multi-fluid is a complex problem especially when the 3D consideration is adopted. However most recently research works are based on the mixture consideration of homogeneous fluid composed by two phases of liquid and vapor, which is also used by our model.

Minimizing the nuisance of cavitation is a great challenge in the design phase of a marine propeller. For efficiency reasons, the propeller usually needs to be operated in cavitating conditions but one still needs to avoid the effects of vibrations, noise and erosion. However, cavitation is a complex phenomenon not yet neither reliably assessable nor fully understood. Experimental observations can only give a part of the answer due to the obvious limitations in the measurement techniques; one example is measuring reentrant jets and internal flow, where flow features are hidden for optical measurement techniques by the cavity itself. Standard simulation tools used in design typically include potential flow solvers, lifting surface or boundary element approaches, with strict theoretical limits on cavitation modeling that only in the hands of an experienced designer may give satisfactory propeller designs. Adding to the challenge is a lack of theoretical knowledge of the physical mechanisms leading to harmful cavitation and thus how to modify a design if some form of nuisance is detected.

The transport equation models of cavitation suggested by Alajbegovic, Grogger et Philipp [1] and Yuan, et al. [2], use the simplistic Rayleigh model. This model describes the limiting case of inertia-controlled growth of a spherical bubble in a liquid under a step variation in pressure of the surrounding liquid. However, this model cannot accurately describe bubble collapse and neglects a number of effects, which determine the behavior of cavitation bubbles.

Recently, progress has been made in the development of numerical models for calculation of cavitation flows. Though the models may differ in terms of realization (using the single-fluid or multi-fluid frame-work, the Eulerian- Eulerian or Eulerian- Lagrangian approaches), all of them are empirical to a certain level.

Modeling of cavitation flow as a multi-fluid is a complex problem which does not lead to satisfying results especially when the cavitation is modeled as 3D. Therefore in the engineering practice cavitation flow is often modeled as a single-fluid, where the cavitation area is handled as an area with the pressure lower then the vapour pressure. This approach always leads to the result, and the requirement of computer time is many times lower in comparison with multi-phase flow models. Moreover the steady solution of multiphase flow model may not be found at all due to the unsteady nature of cavitation flow.

Significant progress has been achieved recently in the development of homogeneous-mixture (single-fluid) models for the simulation of three-dimensional transient cavitating flows (Chen and Heister [3] [4]; Kunz, Boger and Stinebring [5]; Ahuja, Hosangadi et Arunajatesan [6]; Yuan, et al. [2]; Singhal, et al.[7]; Kubota, Kato et

Yamaguchi [8]). These models allow single-fluid solvers to be applied to the conservation equations for the mixture, without increase in computational cost due to the increase in the number of conservation equations when applying the multi-fluid flow concept.

The present work is an investigation to develop a relevant physical model to simulate the cavitating flow. The main goal is the development of computational methodologies which can provide detailed description of the numerical set up for modeling and simulation the cavitation with the CFD code.

MATHEMATICAL FORMULATION

To simulate cavitating flows, the two phases, liquid and vapour, need to be represented in the problem, as well as the phase transition mechanism between the two. Here, we consider a one fluid, single-fluid (mixture), introduced through the local vapour volume fraction and having the spatial and temporal variation of the vapour fraction described by a transport equation including source terms for the mass transfer rate between the phases. The numerical model solves the Reynolds averaged Navier-Stockes equations, coupled with a localized vapour transport model for predicting cavitation.

The fundamental equations governing the flow are taken with incompressible fluid case as given by the Navier-Stockes equations, which control the transport of momentum within the fluid (2) in addition to the mass conservation constraint (1):

$$\frac{\partial \rho_m}{\partial t} + \frac{\partial \left(\rho_m u_j\right)}{\partial x_j} = 0 \tag{1}$$

$$\frac{\partial \left(\rho_m u_i\right)}{\partial t} + \frac{\partial \left(\rho_m u_j u_i\right)}{\partial x_j} = -\frac{\partial p}{\partial x_i} + \frac{\partial}{\partial x_j}\left[\left(\mu_m + \mu_t\right)\left(\frac{\partial u_i}{\partial x_j} + \frac{\partial u_j}{\partial x_i}\right)\right] \tag{2}$$

The effective density and viscosity of the mixture are given (3) and (4) respectively :

$$\rho_m = \rho_v \alpha + (1-\alpha)\rho_l \tag{3}$$

$$\mu_m = \mu_v \alpha + (1-\alpha)\mu_l \tag{4}$$

Where α is the vapor fraction (α ➔1: vapor and for α ➔0: liquid). The density and the viscosity of liquid and vapor are assumed to be constant. To compute the volume fraction we need a closure model, the distribution of values α was obtained on each cell of the computational domain will guide the attendance rate of the vapour.

Vapour Fraction Transport Equation

The transport equation for the vapour scalar fraction α is given by:

$$\frac{\partial \rho_v \alpha}{\partial t} + \frac{\partial(\rho_v \alpha u_j)}{\partial x_j} = \dot{m}^+ + \dot{m}^-$$

This transport equation of volume fraction of vapour, with appropriate source terms to regulate the mass transfer between phases (liquid/vapor), is solved.

The Proposed Model Formulation

a. This work deals with a numerical simulation of cavitation process around a hydrofoil. The numerical approach is based on predicting of the collapse process of pocket of vapor due to liquid compressibility. The model based is the Rayleigh Plesset (dynamics of spherical bubbles) integrated in the term source. Theinterface velocity investigated during the collapse process is given by:

$$\dot{R} = -\sqrt{\left(\frac{2}{3}\frac{p-p_v}{\rho_l}\right)\left(\frac{R_0^3}{R^3}-1\right)}.$$

b. finally, the proposed model can be express :

$$\dot{m}^{-} = C_c \alpha^{\frac{1}{6}} (1-\alpha)^2 \frac{\rho_v \rho_l}{\rho_m} \sqrt{\frac{2}{3} \frac{p - p_v}{\rho_l}}; \ \dot{m}^{+} = C_e \alpha^{\frac{2}{3}} (1-\alpha)^{\frac{4}{3}} \frac{\rho_v \rho_l}{\rho_m} \sqrt{\frac{2}{3} \frac{p_v - p}{\rho_l}} \tag{6}$$

with:

$$C_c = -10 R_0^{\frac{3}{2}} n_0^{\frac{5}{6}}; \ C_e = 5\sqrt[3]{n_0} \tag{7}$$

The value of n_0 is a calibrate parameter with experimental results; it is define the bubble density.

NUMERICAL RESULT

Numerical Study

To validate the proposed model, a confrontation to experimental measures and to the numerical models (Yuan et al., Schnerr and Sauer and EOS). The application is a NACA0009 hydrofoil, truncated at 90% of the original chord length. It has the final dimensions of 100 mm of chord length. The 3D test section is modeled by a quasi 2D domain, with three rows of cells in spanwise direction for the numerical domain. The same mesh and numerical setup is used for the computations with the two models.

Geometry

The domain is 9 blocks (See Fig. 1). The boundary conditions are set using a velocity inlet and a pressure at the outlet (the parameter which fixes the cavitation number).

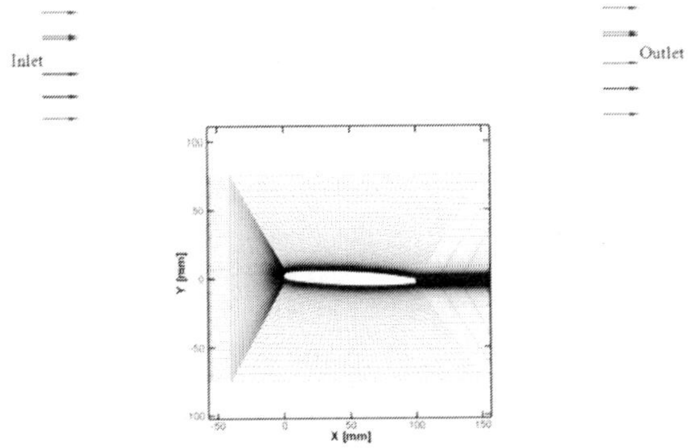

Figure 1: NACA0009 domain grid, y+=10, O-grid structure, with 9 blocks.

The equations of the non-truncated hydrofoil are:

$$0 \le \frac{y}{c} \le 0.45 \qquad \frac{y}{c} = a_0\left(\frac{x}{c}\right)^{\frac{1}{2}} + a_1\left(\frac{x}{c}\right) + a_2\left(\frac{x}{c}\right)^2 + a_3\left(\frac{x}{c}\right)^3$$

$$0.45 \le \frac{y}{c} \le 1 \qquad \frac{y}{c} = b_1\left(1-\frac{x}{c}\right) + b_2\left(1-\frac{x}{c}\right) + b_3\left(1-\frac{x}{c}\right)^3$$

Meshquality

Numerical solutions of fluid flow and heat transfer problems are only approximate solutions. In addition to the errors that might be introduced in the course of the development of the solution algorithm, in programming or setting up the boundary conditions... The discretization approximations introduce errors which decrease as the grid is refined, and that the order of the approximation is a measure of accuracy. However, on a given grid, methods of the same order may produce solution errors which differ by as much as an order of magnitude. Theoretically, the errors in the solution related to the grid must disappear for an increasingly fine mesh [HYPERLINK \l "Fer96"9].

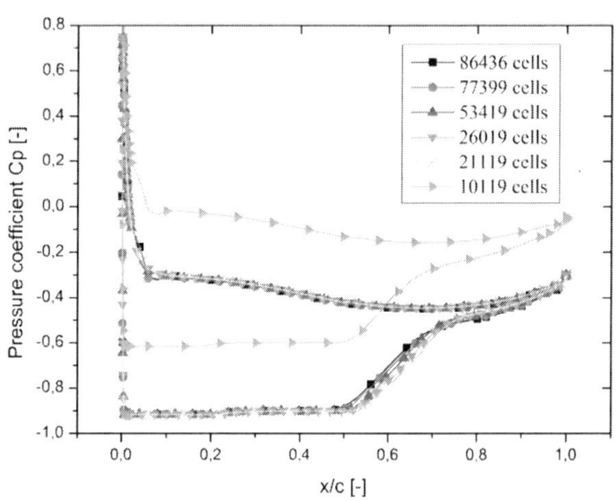

Figure 2: Impact of the quality of the mesh in numerical result, σ=0.4.

The pressure coefficient at σ=0.4 was taken as the parameter to evaluate six grids (See Fig. 2) and determine the influence of the mesh size on the solution. The selected convergence criteria were a maximum residual of 10^{-4}. According to this figure, the grid with 53419 cells is considered to be sufficiently reliable to ensure mesh independence.

Table 1: Numerical parameters

Constants	Mark	Value
Reference velocity@ Inlet	C_{ref}	20 [m s^{-1}]
Length of chord		0.11 [m]
Saturation pressure		3164 [Pa]
Liquiddensity		997 [kg m^{-3}]
Vapourdensity		0.023 [kg m^{-3}]
Reference time		0.0055 [s]
Pressure @ Outlet	P_{out} : calculate as a function of the value of	

Mass, momentum, turbulence and scalar transport equations were solved with Anys CFX which uses a coupled solver, which solves the hydrodynamic equations (for u, v, w, p) as a single system. This solution approach uses a fully implicit discretisation of the equations at any given time step. For steady state problems the time-step behaves like an 'acceleration parameter', to guide the approximate solutions in a physically based manner to a steady-state solution. This reduces the number of iterations required for convergence to a steady state, or to calculate the solution for each time step in a time dependent analysis. For the advection discretization, the High Resolution scheme is employed and the second order backward Euler scheme for the transient term, to assure accurate solution and to reduce the numerical diffusion for the solution [10].

In these calculations turbulence effects were considered using turbulence models, as the k-ε RNG models, with the modification of the turbulent viscosity. To model the flow close to the wall, standard wall-function approach was used. For this model, the used numerical scheme of the flow equations was the segregated implicit solver.

Validation Steady Flow

Turbulence Model Discussion

For steady state flows, there is a priori no difference between the two equations formulations. The use of SST in the case of 2D steady hydrofoil computations instead ofk-ω or k-ε is justified in this way. SST model works by solving a turbulence/frequency-based model (k–ω) at the wall and k-ε in the bulk flow. A blending function ensures a smooth transition between the two models. The SST model performance has been studied in a large number of cases. In a NASA Technical Memorandum, SST was rated the most accurate model for aerodynamic applications [11].

Pressure Coefficient Validation

A validation of the proposed model is described in this section. Firstly, we proceed with a distribution of the pressure coefficient around the surface of the hydrofoil and then we adjust with the profile velocity.

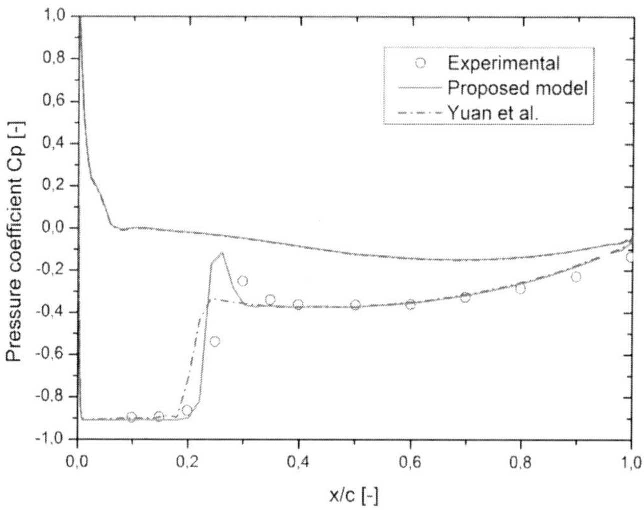

Figure 3: Pressure distribution comparison between proposed model and Yuan et al. model with experimental (measurement reported from [12], on NACA0009, i=2.5°, σ=0.9, = 30 m/s.

Fig. 3 presents the pressure coefficient distribution around the NACA 0009 for the proposed model and the Yuan et al. model with the experimental measurements. The proposed model shows a peak at the closing-pocket (condensation), this is due to the collapse velocity; it tends to compress the vapor pocket (collapse). The results from the proposed model can be satisfied compared to the experimental measurement.

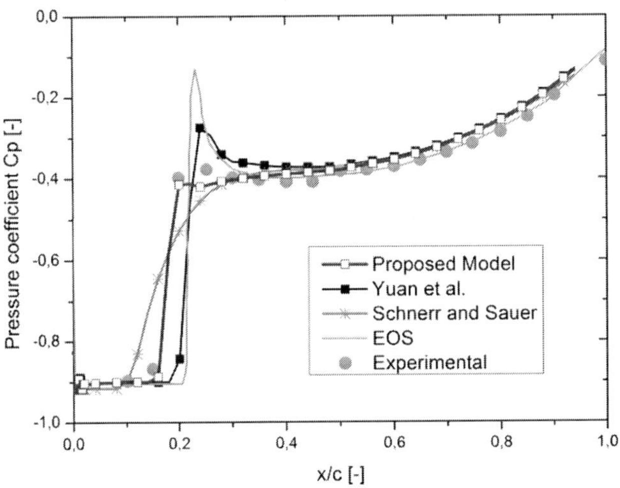

Figure 4: Pressure distribution comparison between proposed model, Yuan et al. model, Schnerr and Sauer model and EOS with experimental (measurement reported from [13]), on NACA0009, i=2.5°, σ=0.9, = 20 m/s.

We remark the good agreement of the numerical values compare to experimental data for $R_0=10^{-6}$ m and $n_0=10^{14}$ bubbles/m^3. These constants are fixed for all this study.

Experimental data concerning the NACA0009 hydrofoil are reported for different cavitation numbers, and compared with the results of the computations in Fig 4.

These figures show satisfactory results of the proposed models in predicting the pressure distribution on the hydrofoil. As expected, the cavity becomes larger with decreasing cavitation number. However the models exhibit very different flow behavior at the cavitation detachment and closure regions. This is the result of dynamic bubble which is implementing in the proposed model. Clearly the model can perfectly predict the cavity of the vapor.

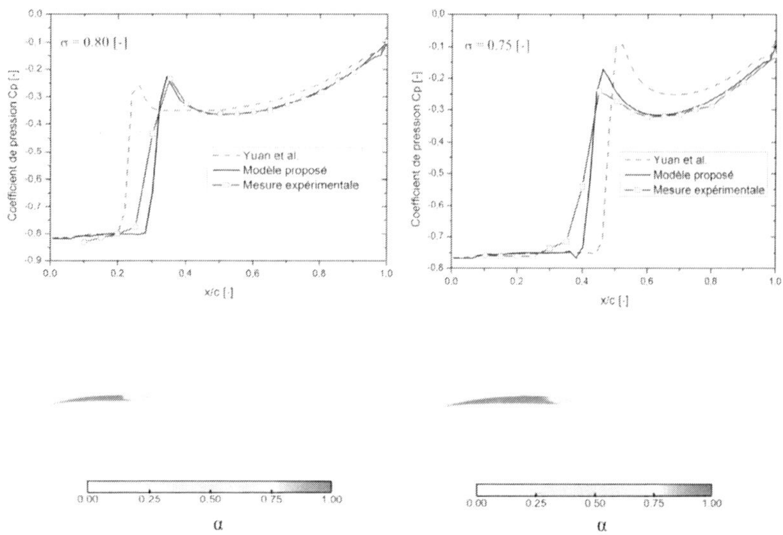

Figure 5: Pressure distribution confrontation between proposed model, Yuan et al. [2] and with experimental (measurement reported from [13]), on NACA0009, U_{ref}= 20 m/s.

Figure 5 show satisfactory results of the proposed model in the pressure distribution on the hydrofoil. As expected, the cavity becomes larger with decreasing cavitation number. However the model exhibit very different flow behavior at the cavitation detachment and closure regions. This is the result of dynamic bubble which is taken account in the proposed model (interface velocity). Clearly the model can perfectly predict the cavity of the vapour.

Velocity Distribution Validation

Figure 6 show the predicted velocity profile of the main flow far from the wall is in good agreement with measurements. In the near-wall and in the wake region, the proposed model tends to reproduce the flow re-entrant jet.

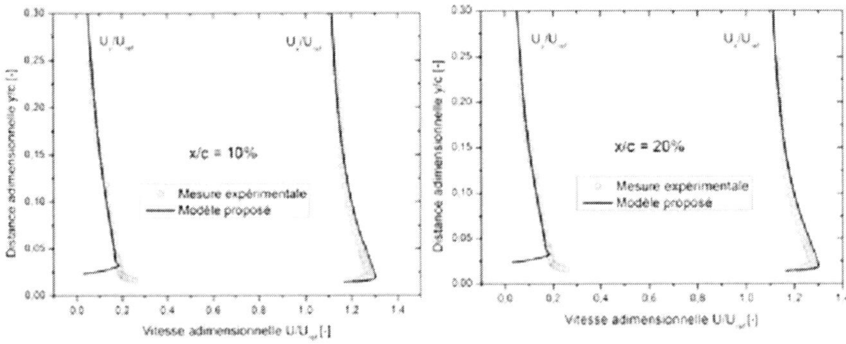

Figure 6: Comparison between computed and measured averaged dimensionless velocity profiles at 10% and 20% of the chord, (chord=100[mm]) (measurement reported from [12]).

Validation Unsteady Flow

2D Configuration (Naca0009)

The vapor cavity is characterized by a thick main cavity and an important shed cavity volume in a disorganized way, such as the cavity can be divided into many small cavities. For one period of cavity creation and collapse, one can remark two distinct life cycles highlighted by the lift and drag signals. Firstly, starting from the maximum cavity length, the closure region is in the small adverse pressure gradient and the reentrant jet is too thin, such as the cavity closure region is continuously broken into small vapor volumes. Secondly, as the reentrant jet reaches the region of the high pressure gradient, the reentrant jet is more important and the whole cavity is extracted and shed downstream. We consider the non-truncated variant of the 2D NACA0009 hydrofoil at high incidence angles. The flow parameters are i=5° and σ=1.2, and reflected by a high pressure gradient over the hydrofoil leading to an unsteady state flow behavior and shedding of large transient cavities. The domain is taken larger than the experimental test section to avoid numerical problem mainly due to reflections on the boundaries.

Cavitating flow are highly sensitive to turbulent fluctuations present in the flow [14] [15]. The effects of modeling turbulence quantities have an enormous impact on the cavitation dynamics and the overall flow structure. The k-ε RNG turbulence model and LES model are adopted in this study.

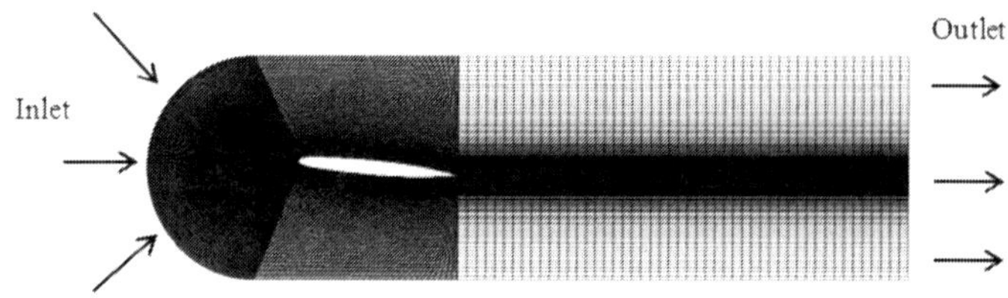

Figure 7: Domain mesh, C-type; $y^{+}=1$; $i=5°$.

Result and Discussion

In this section we present the behavior of the proposed model with LES and the modified turbulence model. Most of characteristics of the unsteady flow is defined as a function of lift and drag coefficients, this coefficient can be expressed as:

$$C_l = \frac{L}{\frac{1}{2}\rho_l C_{ref}^2 A}; \quad C_d = \frac{D}{\frac{1}{2}\rho_l C_{ref}^2 A} \tag{9}$$

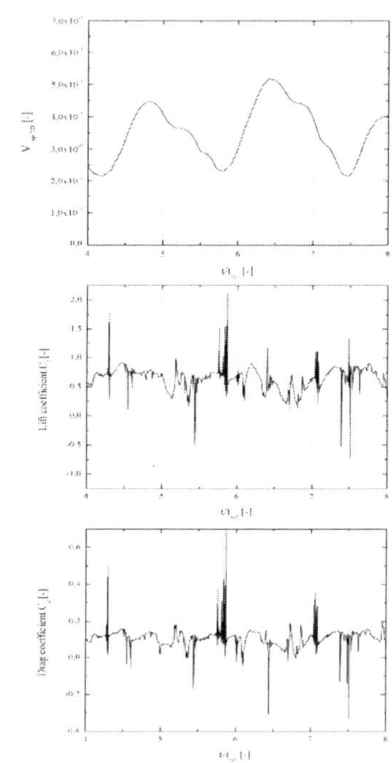

Figure 8: The time history of the lift, the drag coefficient, and total vapour volume for both cavitation numbers σ =1.2, T=0.008 s (the proposed model).

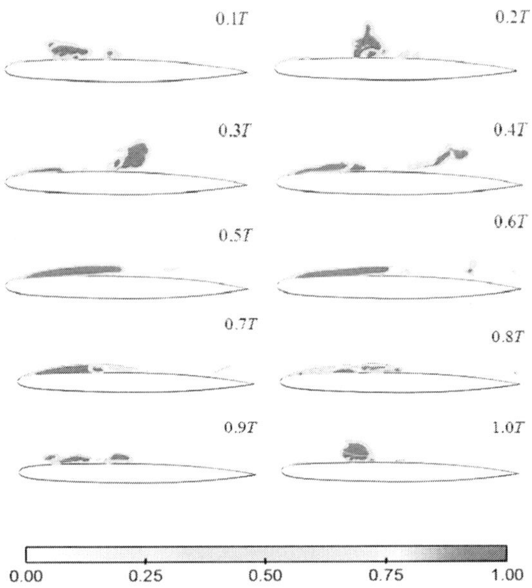

Figure 9: Volume fraction of vapor for σ=1.2, for life cycle (period T=0.01s).

For flows about 2D hydrofoils the dimensionless total vapour volume *Vvap,2D* is defined as :

$$V_{vap,2D} = \frac{1}{c^2}\sum_{i=1}^{N}\alpha_i V_i \qquad (10)$$

where N is the total number of control volumes and α is the volume fraction of vapour in each control volume, with the volume Vi of the fluid in control volume.

The total vapour volume *Vvap*, 2D, defined in equation (10), is a convenient parameter for understanding the transient evolution of the cavitating flow. The total vapour volume is calculated at each time step. After the start-up phase the growth and shedding of the vapour sheet and the collapse of the shed vapour cloud induce a self-oscillatory behavior, which is approximately periodic in time. The graphics of the analyzed variables (Fig.8) show a periodic signal, even if the identified periods are very different from each others. The use of Vvap, 2D is the easiest way to identify the periodicity of the vapour formation and collapse during a life cycle. Furthermore, the time-history of the lift and drag coefficients are compared to those of the total vapour volume to correlate the occurring flow phenomena. The lift and drag coefficient has a cyclic time signal. Even if the periods are pretty clear, the fluctuations in a given period are very different from one to another. This phenomenon is mainly due to the dynamic of the shed cavities which are driven by the main flow-field downstream of the cavity closure. The growth and collapse mechanism of the cavity is driven by a cyclic phenomenon, whereas the dynamic of the cavity, when it is swept away can have very different non reproducible behavior. The vapour can be attached to the wall or far from it. This different ways of cavities shedding have an important influence on the pressure field at the hydrofoil wall, and there by on the lift and drag values.

In this section the cycles illustrated in figures 15 and 16 are considered. The solution for the volume fraction α above the hydrofoil is presented for a number of equidistant time-intervals during the cycle.

The vapour cavity is characterized by a thick main cavity and an important shed cavity volume in a disorganized way, such as the cavity can be divided into many small cavities. For one period of cavity creation and collapse, one can remark two distinct life cycles highlighted by the lift and drag signals.

Firstly, starting from the maximum cavity length, the closure region is in the small adverse pressure gradient and the reentrant jet is too thin, such as the cavity closure region is continuously broken into small vapour volumes. Secondly, as the reentrant jet reaches the region of the high pressure gradient, the reentrant jet is more important and the whole cavity is extracted and shed downstream.

3D Configuration(NACA66MOD)

These excellent results obtained for 2-D cavitating flow confirm the correct assumptions of the proposed numerical model in terms of vaporization and condensation processes, and to verify it's performance with 3-D considerations, we present in the next some numerical results compared to experimental measurement [16] for cavitating flow around hydrofoil NACA66 (Fig 10).

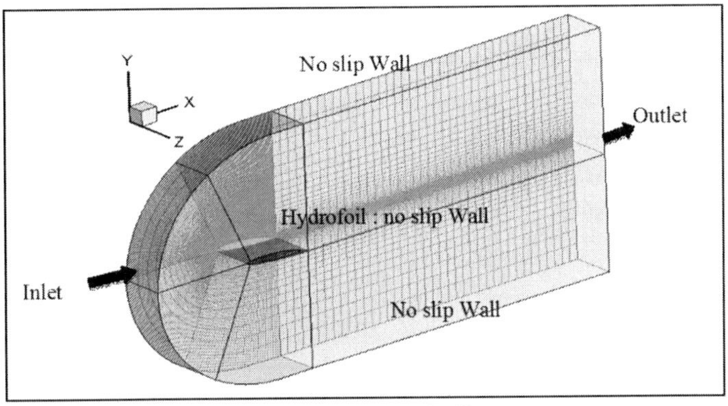

Figure 10: Domain mesh, C-type; $y^{+}=50$; $i=6°$.

In this section, a numerical simulation was performed to further study the performance of the proposed model and unsteady three-dimensional, the choice is justified because of the availability of experimental measurements. The profile used is NACA66 (mod) - 312 a = 0.8 with a string of 0.15m, the width of the profile is 0.19m and is placed at 6 ° incidence relative to the upstream flow, not in an infinite medium -viscous. It has a thickness of approximately 12% and a camber on 2% to 45% and 50% of the leading edge along the rope. Theoretical points of this section have been interpolated by B-spline technique using the software mesh.

The study area is in 3D, it is made larger than the experimental test section to avoid numerical problems mainly due to reflections from the boundary conditions. The estate consists of 156 000 cells, the mesh topology is structured with a C-type conditions with wall on its border, the hydrofoil with the no-slip condition. (see Fig. 8). The steady state solution was used as an initial condition, the turbulence model used is k-ε RNG.

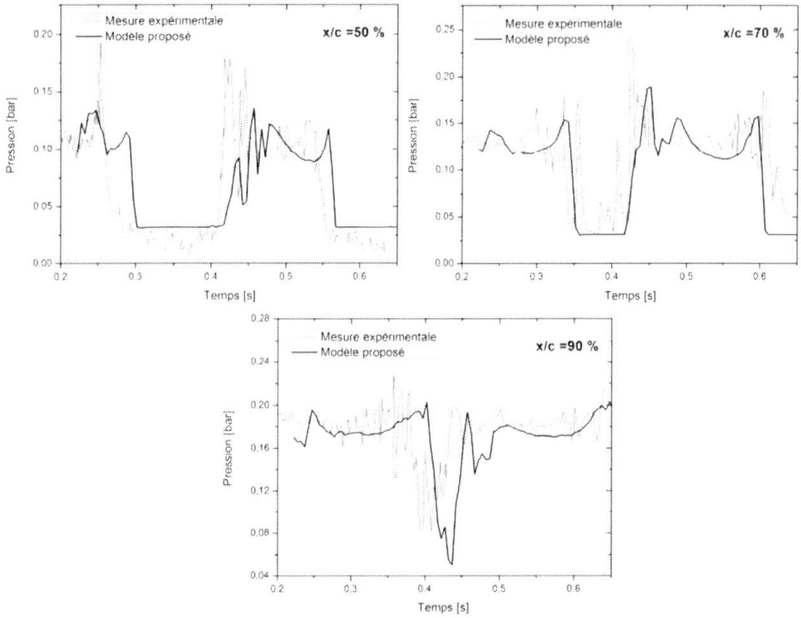

Figure 11: Domain grid and boundary condition at the left, at the right, the validation of the pressure measurement[16] respectively at 50%,70% and 90% of the chord length with the proposed model, *Cref*= 5.3 m/s.

The proposed model was validated quantitatively and qualitatively compared in two-dimensional distribution of pressure coefficient and velocity profiles calculated with experimental measurements. We propose in the following one-dimensional confrontation, the application as we have described above is the NACA66 (mod), the advantage of selecting this application is the availability of measuring pressure along the profile for two cycles of detachment from the pocket of steam. We have reported three points of pressure measurement located respectively 50%, 70% and 90% chord.

The obtained numerical result presented by figure9 show that the model reproduce correctly the typical behaviour of partial cavity with development of re-entrant jet and the periodic shedding of cavitation clouds. Firstly, starting from the maximum cavity length, the closure region is in the small adverse pressure gradient and the re-entrant jet is too thin, such as the cavity closure region is continuously broken in to small vapour volumes. Secondly, as the re-entrant jet reaches the region of the high pressure gradient, the re-entrant jet is more important and the whole cavity is extracted and shed downstream.

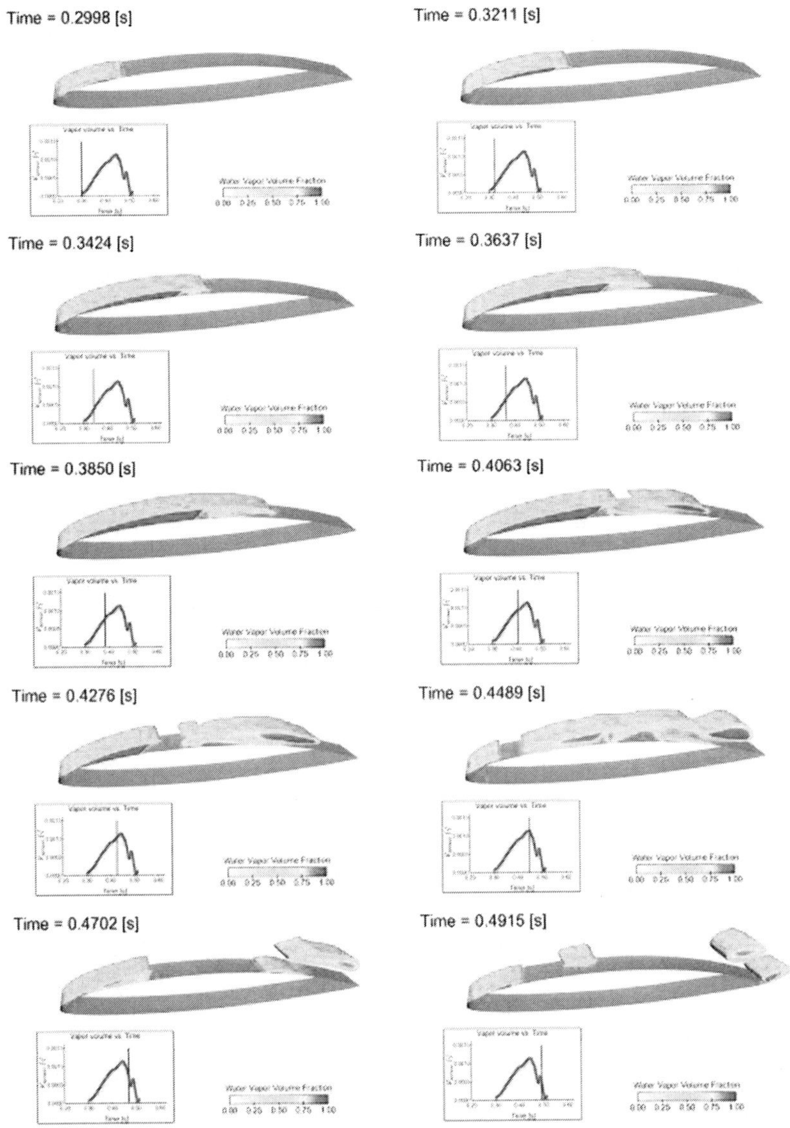

Figure 12: Volume fraction of vapor for σ=1, for life cycle (period T=0.2s).

This different ways of cavities shedding have an important influence on the pressure field at the hydrofoil wall, and there are characterized by the temporal evolution of the lift coefficient Cl compared to experimental result [16].

CONCLUSION

A comprehensive theoretical approach is done, and detailed formulations of the proposed model are presented. Influence of the numerical parameter has been widely studied. A comparative study is made for the steady flow. Good agreement with measurements was obtained for the proposed model. We have shown the importance of liquid compressibility on the pocket of vapour and how it is modeling in the proposed model. Computations with the proposed model is compared with experimental data and other numerical models, it shows the ability of the model to reproduce the steady-state developed cavitation flow fields. Finally, the unsteady behaviour of the cavitating flow depends strongly on the turbulence model, a k-ε RNG model is adopted in this steady.

REFERENCES

1. A Alajbegovic, H. A Grogger, and H Philipp, Calculation of transient cavitation in nozzle using the two-fluid model," in Proc. ILASS-Americas'99 AnnualConf., 1999373377
2. W Yuan, J Sauer, H. G Schnerr, Mec. Ind., 20012383394
3. Y Chen, and S. D Heiste, Modelling Hydrodynamic Nonequilibrium in Cavitating Flows," in Tr. ASME Journal Fluids Eng, 1181996172178
4. Y Chen, and S. D Heister, Two-phase modelling of cavitated flows," in In ASME Cavitation and Multiphase Forum, 24Reno, Nevada, 1995799806
5. R. F Kunz, D. A Boger, and D. R Stinebring, A preconditioned Navier-Stokes method for two-phase flows with application to cavitation predictionComputer & Fluids, 298498752000
6. V Ahuja, A Hosangadi, and S Arunajatesan, Simulations of caviting flows using hybrid unstructured meshes," J. Fluids Engng, 1233313392001

7. A. K Singhal, M. M Athavale, H Li, and Yu Jiang, "Mathematical basis and validation of the full cavitation model1242002617624
8. A Kubota, H Kato, and H Yamaguchi, A new modelling of cavitating flows: a numerical study of unsteady cavitation on a hydrofoil sectionJournal of Fluid Mechanics2401992
9. J. H Ferziger, and M Peric, Computational Methods for Fluid DynamicsBerlin, Germany: Springer, 1996
10. ANSYS CFX,201069
11. J. E Bardina, P. G Huang, and T. J Coakley, Turbulence Modeling Validation, Testing and Development," NASA Technical Memorandum 110446, AIAA, 1997
12. P. H Dupont, Etude de la dynamique d'une poche de cavitation partielle en vue de la prediction de l'érosion dans les turbomachines hydrauliques," 1993
13. YoucefAitBouziadphysical modelling of leading edge cavitation: computational methodologies and application to hydraulic machinery.: Lausanne, EPFL, 2006
14. O Coutier-delgosha, R Fortes-patella, J. L Reboud, M Hofmann, and B Stoffel, Experimental and Numerical Studies in a Centrifugal Pump with Two-Dimensional curved Blades in Cavitating condition125no. Transactions of the ASME, 9709782003
15. J. L Reboud, O Coutier-delgosha, B Pouffary, and R Fortes-patella, Numerical Simulation of Unsteady Cavitating Flows: some Applications and Open Problems," in In Fifth International Symposium on Cavitation, 2003
16. J. B Leroux, J. A Astolfi, and J. Y Billard, EXPERIMENTAL STUDY OF UNSTEADY AND UNSTABLE PARTIAL CAVITATION," in 9éme journée de l'hydrodynamique, 2003

CITATION

Hatem Kanfoudi, Hedi Lamloumi and Ridha Zgolli (2012). Numerical Investigation for Steady and Unsteady Cavitating Flows, Advances in Modeling of Fluid Dynamics, Dr. Chaoqun Liu (Ed.), ISBN: 978-953-51-0834-4, InTech, DOI: 10.5772/48421.

CHAPTER 2

Laminar Motion of the Incompressible Fluids in Self-Acting Thrust Bearings with Spiral Grooves

Cornel Velescu[1] *and NicolaeCalin Popa*[2]

[1]Hydraulic Machinery Department, Mechanical Faculty, "Politehnica" University of Timisoara, Boulevard MihaiViteazul No. 1, 300222 Timisoara, Romania

[2]Center for Advanced and Fundamental Technical Research (CAFTR), Romanian Academy-Timisoara Branch, Boulevard MihaiViteazul No. 24, 300223 Timisoara, Romania

ABSTRACT

We analyze the laminar motion of incompressible fluids in self-acting thrust bearings with spiral grooves with inner or external pumping. The purpose of the study is to find some mathematical relations useful to approach the theoretical functionality of these bearings having magnetic controllable fluids as incompressible fluids, in the presence of a controllable magnetic field. This theoretical study approaches the permanent motion regime. To validate the theoretical results, we compare them to some experimental results presented in previous papers. The laminar motion of incompressible fluids in bearings is described by the fundamental equations of fluid dynamics. We developed and particularized these equations by taking into consideration the geometrical and functional characteristics of these hydrodynamic bearings. Through the integration of the differential equation, we determined the pressure and speed distributions in bearings with length in the "pumping" direction. These pressure and speed distributions offer important information, both quantitative (concerning the bearing performances) and qualitative (evidence of the viscous-inertial effects, the fluid compressibility, etc.), for the laminar and permanent motion regime.

INTRODUCTION

Incompressible fluid motion in "self-acting thrust bearings with spiral grooves" (SATBESPIG) is a complex motion in thin layers [1–5] bounded by two solid surfaces in relative rotation one to another. The incompressible fluid motion in thin layers has as mathematical consequence the simplification and particularization of the general equations of motion. We made this simplification of the general equations of motion by neglecting the terms smaller by more than one order of magnitude [1–7] than the other terms. Starting from this observation, the paper approaches the theoretical study of incompressible fluid motion in SATBESPIG with inner and exterior pumping, working in the laminar regime.

Some authors have developed comparative theoretical and experimental studies concerning different aspects of spiral groove bearings functioning in the laminar regime, with equations for their functioning using magnetic fluids, cavitation phenomenon, and the geometry of the bearings with spiral/axial multiples grooves.

In [8, 9], the author studied the static and dynamic performances of the bearings with spiral grooves and inner and exterior pumping, working with compressed air. A comparative study of the theoretical and experimental results was presented.

The diversification and optimization of the geometry of the bearings with spiral or axial multiples of grooves were approached in [10–12]. In [13], the author analyzed the cavitation phenomenon theoretically and experimentally in bearings with spiral grooves working in mineral oil. The unfavorable consequences of the cavitation were evaluated.

We consider the theoretical results as a first step in our intention to approach the functionality of the hydrodynamic bearings in general and of the thrust bearings with spiral grooves in particular in the stationary/nonstationary turbulent regime. The viscous compressible/incompressible fluid motion can be approached by taking into consideration the thermo-viscous-inertial phenomena ensemble that exists in bearings.

CONSTRUCTIVE GEOMETRIC AND CINEMATIC ELEMENTS OF THE SATBESPIG

The SATBESPIG (Figures 1 and 2) are special hydrodynamic bearings that have the specific effect of "auto pumping" [1–3, 5, 7], which is why we are interested in the incompressible fluid motion in the laminar and permanent regime and, taking into consideration the influence of the inertial forces, in the length of the spiral groove (pumping direction ψ from Figures 3, 4, and 5). In the literature [4, 7, 10, 12–15], there are only a few studies and a small amount of theoretical research concerning this subject. These studies approach the incompressible fluid motion only in the radial direction (r) without taking into consideration the effects of the inertial forces.

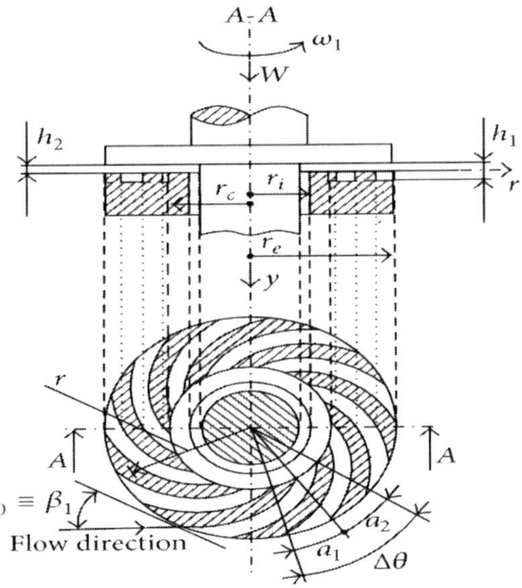

Figure 1: SATBESPIG and inner pumping.

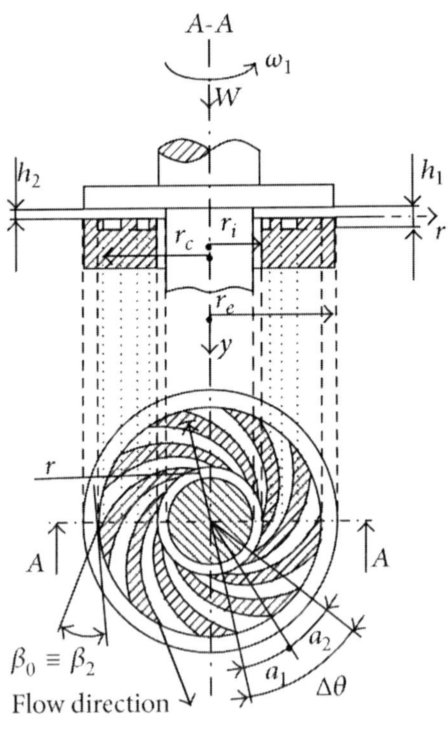

Figure 2: SATBESPIG and external pumping.

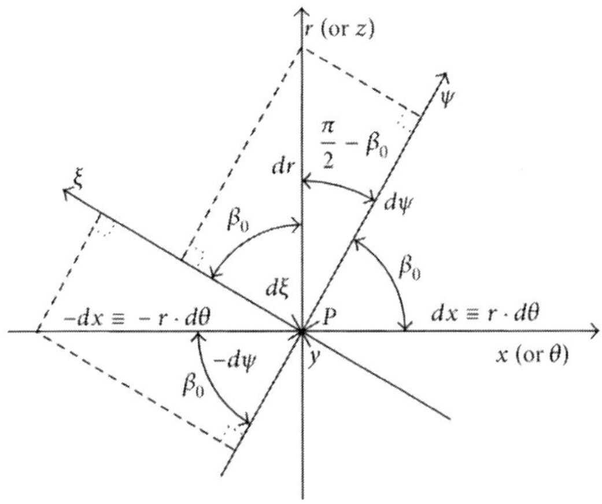

Figure 3: Relations between the general and local coordinate systems (in an arbitrary point, P). At point P, the coordinate ξ is normal to the ψ direction of the spiral channel.

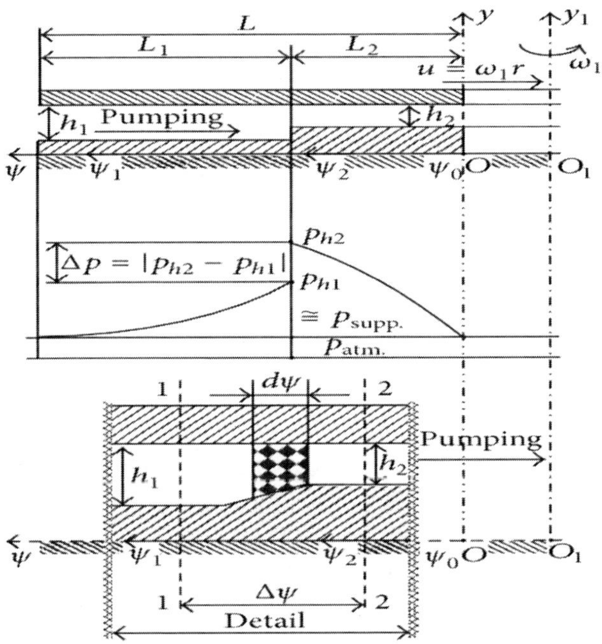

$$\psi_0 \leq \psi_2 \leq L_2 = L - L_1;\ L - L_1 = L_2 \leq \psi_1 \leq L$$

Figure 4: Radial step of the inner pumping bearing.

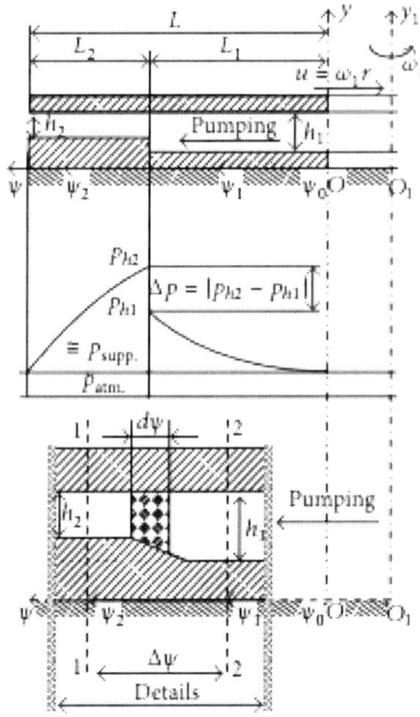

$\psi_0 \leq \psi_1 \leq L_1 = L - L_2;\ L - L_2 = L_1 \leq \psi_2 \leq L$

Figure 5: Radial step of the external pumping bearing.

In Figures1 and 2, (i) a_1 is the width of the spiral channel, measured on the circle arc contour; (ii) a_2 is the width of the spiral threshold, measured on the circle arc contour; (iii) $\Delta\theta$ is the center angle corresponding to a channel—threshold pair; (iv) β_0 is the generator angle of the logarithmic spiral, which describes the form of the cannels; (v)β_1 andβ_2 are the input and output angles, respectively, in and from the channel (according to the accepted notations from the general study of the hydraulic machineries); (vi) ω_1 is the bearing angular rotation speed; (vii) r_i andr_e are the inner and external radius of the bearing, respectively; (viii) r_c is the radius marking the zone of the spiral channels; (ix) r is the current radius of the bearing; (x) h_1 is the lubrication film height over the bearing channels; (xi)h_2 is the lubrication film height over the bearing thresholds; (xii)W is the bearing load (the weight); and (xiii) by definition, α=$a1/(a_1 + a_2)$. With the geometrical dimensions in Figure 1(or Figure 2) we can write $\Delta\theta = 2\pi/n_p$, where n_p is the number of pairschannel—thresholds, $a1 = \alpha r\Delta\theta$, $a_2 = (1 - \alpha)r\Delta\theta$, a=$a_1 + a_2 = r\Delta\theta$, and$\omega_2 = 0$ (the grooved surface is fixed).

COORDINATE SYSTEMS, CONTROL VOLUME, SPEED DISTRIBUTIONS, AND MASS FLOW RATE IN LAMINAR REGIME

To study the mathematical model, the coordinate systems and the control volume must be determined so as to define the speed distributions and the fluid mass flow rate [3, 5, 15]. The control volume (Vol) is the volume between the bearing surfaces, between the one channel surface and one consecutive threshold surface of the stator and the horizontal bottom surface of the rotor (Figures 1 and 2). In Figure 3, we show the general (y, r, θ) and local (ψ, y, ξ) coordinate systems used for the motion study.

Next, we start from the fact that the incompressible fluid laminar motion, between the two quasiparallel surfaces of the bearing, is described by the following speed profiles:

$$u(y) \cong -\frac{1}{2\eta}\frac{1}{r}\frac{\partial p}{\partial \theta}y(h-y)+\frac{y}{h}\omega_1 r, \tag{1a}$$

$$\upsilon_n(y) \cong 0, \tag{1b}$$

$$w(y) \cong -\frac{1}{2\eta}\frac{\partial p}{\partial r}y(h-y), \tag{1c}$$

$$\upsilon_\psi(y) \cong \frac{1}{2\eta r\sin\beta_0}\frac{\partial p}{\partial \psi}y(y-h)+\frac{y}{h}\omega_1 r\cos\beta_0, \tag{1d}$$

$$\upsilon_\xi(y) \cong \frac{1}{2\eta r\sin\beta_0}\frac{\partial p}{\partial \xi}y(y-h)-\frac{y}{h}\omega_1 r\sin\beta_0. \tag{1e}$$

In (1a)–(1e), (i)$u(y)$is the fluid speed component in the x (or θ) direction, (ii) V$n(y)$is the fluid speed component in the normal direction$y$, (iii)$w(y)$is the fluid speed component in the z (or r) direction, (iv) V$\psi(y)$is the fluid speed component in theψcurb direction, (v)V$\xi(y)$is the fluid speed component in the ξ direction, (vi) p is the pressure in the fluid, (vii) $h$is the height of the lubrication film, and (viii) η is the fluid dynamical viscosity.

The speed distributions, given by (1a)–(1e), are typical for the noninertial motion case initially considered, meaning that these are parabolic speed profiles [3, 5, 6]. Observing the geometry of the bearings in Figures 1 and 2, it is possible to establish some functional mathematical relations between the coordinates. Thus, the following operational expressions canbe found [3, 5]:

$$\frac{\partial}{\partial r} = \cos\beta_0 \frac{\partial}{\partial \xi} + \sin\beta_0 \frac{\partial}{\partial \psi}, \tag{2a}$$

$$\frac{\partial}{\partial x} \equiv \frac{1}{r}\frac{\partial}{\partial \theta} = -\sin\beta_0 \frac{\partial}{\partial \xi} + \cos\beta_0 \frac{\partial}{\partial \psi}, \tag{2b}$$

$$\frac{\partial}{\partial \xi} = \cos\beta_0 \frac{\partial}{\partial r} - \sin\beta_0 \frac{1}{r}\frac{\partial}{\partial \theta}, \tag{2c}$$

$$\frac{\partial}{\partial \psi} = \sin\beta_0 \frac{\partial}{\partial r} + \cos\beta_0 \frac{1}{r}\frac{\partial}{\partial \theta}. \tag{2d}$$

Using relation (1e), the fluid mass flow rate in the pumping direction ψ (Figures 4 and 5), further denoted by Mψ, can be expressed by the integral representation [3, 5]:

$$\dot{\mathcal{M}}_\psi \cong \int_\theta^{\theta+\Delta\theta} r_0 \sin\beta_0 d\theta \int_0^h \rho v_\psi(y)\, dy, \tag{3}$$

whereρ is the fluid density and $r0$ is the "reference" radius ($r0 \in [ri,...,re]$, and further $r0$ will be denoted by r).

DIFFERENTIAL EQUATION FOR PRESSURE DISTRIBUTION IN THE ψ DIRECTION

Observing the bearing geometry (Figures1 and 2), we admit that the angle $\Delta\theta$ is infinitely small, meaning that there existan infinite number of spiral channels. Given the relations (1e) and (3), the physical natural condition is that the mass flowRateM_ψ is constant. With these conditions we obtain

$$\dot{\mathcal{M}}_\psi \cong \Delta\theta \left[C\frac{\partial p}{\partial \psi} + Dr^2 \right], \tag{4}$$

Where

$$C = -\frac{\rho}{12\eta}\left[\alpha h_1^3 + (1-\alpha) h_2^3\right], \tag{5a}$$

$$D = \frac{\rho}{2}\omega_1 \sin\beta_0 \cos\beta_0 \left[\alpha h_1 + (1-\alpha) h_2\right]. \tag{5b}$$

The fluid mass conservation in the ψ direction can be expressed as follows:

$$\frac{\partial}{\partial \tau}(m) + \frac{\partial}{\partial \psi}\left(\dot{\mathcal{M}}_{\psi}\right)\Delta\psi \cong 0, \tag{6}$$

Where τ is the time, and the fluid mass m, contained in the control volume Vol, is given by the relation

$$m \cong \Delta r \Delta\theta \rho r \left[\alpha h_1 + (1-\alpha) h_2\right]. \tag{7}$$

Or, using relations (4) and (7), (6) becomes

$$\frac{\partial}{\partial \tau}\{[\alpha h_1 + (1-\alpha) h_2]\rho\} + \sin^2\beta_0 \frac{1}{\psi}\frac{\partial}{\partial\psi}\left\{C\frac{\partial p}{\partial\psi} + D\frac{1}{\sin^2\beta_0}\psi^2\right\} \cong 0. \tag{8}$$

For the permanent motion regime, (8) becomes

$$\frac{1}{\psi}\frac{\partial}{\partial\psi}\left\{C\frac{\partial p}{\partial\psi} + \frac{1}{\sin^2\beta_0}D\psi^2\right\} \cong 0. \tag{9}$$

Using non-dimensional variables [1, 3–5, 7, 10, 14, 15], (8) can be written as

$$\frac{\partial}{\partial \zeta}\left[PH^3\left(K_1\zeta\frac{\partial P}{\partial \zeta}+K_2\Omega\Lambda\zeta^2H^{-2}\right)\right]-\sigma\zeta\frac{\partial}{\partial t}\left[PHK_3\right]\cong 0. \tag{10}$$

For the stationary motion regime, (10) becomes [1, 2]

$$\frac{\partial}{\partial \zeta}\left[P\left(K_1\zeta\frac{\partial P}{\partial \zeta}+K_2\Omega\Lambda\zeta^2\right)\right]\cong 0. \tag{11}$$

INTEGRATION OF THE DIFFERENTIAL EQUATION OF THE PRESSURE DISTRIBUTION IN THE ψ DIRECTION

In theψdirection, the fluid film$h$varies rapidly from$h1$ toh_2,at the frontier $r\cong rc$ (Figures 4 and 5).The existing radial step with length in the pumping direction ψ produces a pressure jump from$ph1$ to$ph2$. This pressure jump has different values as a function of (i) the flow regime through bearing (laminar, transition, or turbulent regime), (ii) the value of the rapport h_1/h_2, (iii) the fact that we take (or not) into consideration the inertial forces, and (iv) the fluid type [2, 3, 5,16,17].

Integrating differential equation (9) and observing the limit conditions for pressures [3, 5] and the notations from Figures4and5 ($p$supp. is the supply pressure of the lubrication fluid and $p$atm. is the atmospheric pressure), we obtain the mathematical relations for the pressure distributions in the SATBESPIG:

$$\begin{aligned} p(\psi) &\cong p_{h2}+\frac{\psi-(L_2+\psi_0)}{\psi_0-(L_2+\psi_0)} \\ &\times\left\{\left(p_{\text{supp.}}-p_{h2}\right)-\frac{2\eta\cos\beta_0\omega_1}{h_2^2\sin\beta_0}\left[\psi_0^3-(L_2+\psi_0)^3\right]\right\} \\ &+\frac{2\eta\cos\beta_0\omega_1}{h_2^2\sin\beta_0}\cdot\left[\psi^3-(L_2+\psi_0)^3\right]. \end{aligned} \tag{12}$$

Relation (12) presents the pressure distribution in the laminar and permanent flow regime in the smooth region of the inner pumping bearing surface, where h=$h2$. Consider.

$$
\begin{aligned}
p(\psi) & \\
\cong p_{\text{supp.}} &+ \frac{\psi - (L + \psi_0)}{(L_2 + \psi_0) - (L + \psi_0)} \\
&\times \left\{ \left(p_{h1} - p_{\text{supp.}} \right) - \frac{2\eta \cos \beta_0 \omega_1 \left[\alpha h_1 + (1 - \alpha) h_2 \right]}{\sin \beta_0 \left[\alpha h_1^3 + (1 - \alpha) h_2^3 \right]} \right. \\
&\left. \times \left[(L_2 + \psi_0)^3 - (L + \psi_0)^3 \right] \right\} \\
&+ \frac{2\eta \cos \beta_0 \omega_1 \left[\alpha h_1 + (1 - \alpha) h_2 \right]}{\sin \beta_0 \left[\alpha h_1^3 + (1 - \alpha) h_2^3 \right]} \left[\psi^3 - (L + \psi_0)^3 \right].
\end{aligned} \tag{13}
$$

Relation (13) presents the pressure distribution in the region with spiral channels of the inner pumping bearing surface, where h=h_1. In a similar way, for the SATBESPIG with exterior pumping (Figure 5), we obtain

$$
\begin{aligned}
p(\psi) & \\
\cong p_{h2} &+ \frac{\psi - (L_1 + \psi_0)}{(L + \psi_0) - (L_1 + \psi_0)} \\
&\times \left\{ \left(p_{\text{supp.}} - p_{h2} \right) - \frac{2\eta \cos \beta_0 \omega_1}{h_2^2 \sin \beta_0} \left[(L + \psi_0)^3 - (L_1 + \psi_0)^3 \right] \right\} \\
&+ \frac{2\eta \cos \beta_0 \omega_1}{h_2^2 \sin \beta_0} \left[\psi^3 - (L_1 + \psi_0)^3 \right].
\end{aligned} \tag{14}
$$

Relation (14) presents the pressure distribution in the laminar and permanent flow regime in the smooth region of the external pumping bearing surface, where h=h_2. Consider

$$p(\psi) \cong p_{\text{supp.}} + \frac{\psi - \psi_0}{(L_1 + \psi_0) - \psi_0} \times \left\{ \left(p_{h1} - p_{\text{supp.}} \right) - \frac{2\eta \cos \beta_0 \omega_1 \left[\alpha h_1 + (1-\alpha) h_2 \right]}{\sin \beta_0 \left[\alpha h_1^3 + (1-\alpha) h_2^3 \right]} \times \left[(L_1 + \psi_0)^3 - \psi_0^3 \right] \right\} + \frac{2\eta \cos \beta_0 \omega_1 \left[\alpha h_1 + (1-\alpha) h_2 \right]}{\sin \beta_0 \left[\alpha h_1^3 + (1-\alpha) h_2^3 \right]} \left[\psi^3 - \psi_0^3 \right]. \tag{15}$$

Relation (15) presents the pressure distribution in the laminar and permanent motion regime in the spiral grooves region of the bearing surface with exterior pumping, where $h=h_1$. In the relations (12)–(15), all the constants are known (for a designed and realized SATBESPIG), the exception being the extreme pressures ph_1 and ph_2. If we do not take into consideration the influence of the inertial forces, then the pressures ph_1 and ph_2 are equal. Thus, $ph_1 \equiv ph_2$, $\Delta p = ph_2 - ph_1 = 0$.

CALCULUS RELATIONS FOR THE EXTREME PRESSURES ph_1 AND ph_2

To express the pressure distribution in the ψ direction and the extreme pressuresph_1 andph_2, we must analyze the liquid motion in the fluid film existing between the quasi parallel surfaces of the SATBESPIG [1–3, 5, 7]. On the other hand, the inertial effects (which considerably influence the extreme pressures ph_1 and ph_2) exist on all the surfaces of the SATBESPIG in the ψ direction, but the maximum effect is concentrated in the zone of the radial step, $r \cong rc$ (Figures 4 and 5) [2, 3, 5,16,17].

Some theoretical results concerning the motion of liquids in similar bearings with the consideration of the influence of inertial forces have been presented in the literature [3, 5–7, 14]. It is possible to demonstrate [3, 5] that, for the case of the stationary motion regime and when only the smooth surface is in a rotation with $n1 =$

constant [rot/min], the equation that describes the viscous fluid motion inψdirection is

$$\frac{\rho}{\psi}\frac{d}{d\psi}\left(\frac{\alpha_0 Q_\psi^2}{h\psi^2\Delta\theta^2} - \frac{\gamma Q_\psi \omega_1}{\Delta\theta}\frac{\cos\beta_0}{\sin\beta_0} + \beta\omega_1^2\psi^2\frac{\cos^2\beta_0}{\sin^2\beta_0}h\right)$$
$$+ \frac{2\rho\delta Q_\psi\omega_1 \sin\beta_0}{\psi^2\Delta\theta\cos\beta_0} - 3\rho\beta\omega_1^2 h + \frac{\rho\alpha_0 Q_\psi^2}{h\psi^4\Delta\theta^2} \tag{16}$$
$$- \frac{\rho\gamma Q_\psi\omega_1\cos\beta_0}{\psi^2\Delta\theta\sin\beta_0} + \rho\beta\omega_1^2\frac{\cos^2\beta_0}{\sin^2\beta_0}h + \frac{1}{\psi}\frac{dp}{d\psi}h$$
$$+ \frac{12\eta}{h}\left(\frac{Q_\psi}{h\psi\Delta\theta} - \frac{\omega_1\psi\cos\beta_0}{2\sin\beta_0}\right) \cong 0,$$

Where

$$Q_\psi = U_{av.}h \cong -\frac{h^3}{12\eta r\sin\beta_0}\frac{\partial p}{\partial\psi} \tag{17a}$$
$$+ \frac{h}{2}\omega_1 r\cos\beta_0 = \text{const. (volumic flow rate)},$$

$$U_{av.}$$
$$= \frac{1}{h}\int_0^h v_\psi(y)\cdot dy \tag{17b}$$
$$= \text{average speed in the fluid film by the curb direction } \psi,$$

$$\alpha_0 = \frac{6}{5}, \quad \beta = \frac{2}{15}, \quad \gamma = \frac{1}{5}, \quad \delta = \frac{1}{10}. \tag{17c}$$

The constructive angle of spiral groove is $\beta0 = 17\circ$ [3, 5,14]. Calculating the derivative of the relation (16) versus the variable ψ and taking into consideration the constants from above, the following expression can be found:

$$\frac{dp}{d\psi} \cong \left(\frac{\rho\alpha_0 Q_\psi^2}{h^3\psi^2\Delta\theta^2} - \frac{\rho\beta\omega_1^2\psi^2}{h}\frac{\cos^2\beta_0}{\sin^2\beta_0} \right)\frac{dh}{d\psi} + \frac{\rho\alpha_0 Q_\psi^2}{h^2\psi^3\Delta\theta^2}$$
$$+ 3\rho\beta\omega_1^2\psi\left(1 - \frac{\cos^2\beta_0}{\sin^2\beta_0}\right) - \frac{2\rho\delta Q_\psi\omega_1\sin\beta_0}{h\psi\Delta\theta\cos\beta_0} \tag{18}$$
$$+ \frac{\rho\gamma Q_\psi\omega_1\cos\beta_0}{h\psi\Delta\theta\sin\beta_0} - \frac{12\eta Q_\psi}{h^3\Delta\theta} + \frac{6\eta\omega_1\psi^2\cos\beta_0}{h^2\sin\beta_0}.$$

We obtain a similar but more precise relation to (18) byintroducing a supplementary coercive term [1, 2, 4, 9]. In this case, relation (18) becomes

$$\frac{dp}{d\psi} \cong \left[\frac{\rho Q_\psi^2}{h^3\psi^2\Delta\theta^2}(\alpha_0 + \varepsilon\psi) \right]\frac{dh}{d\psi} - \frac{\rho\beta\omega_1^2\psi^2}{h}\frac{\cos^2\beta_0}{\sin^2\beta_0}\frac{dh}{d\psi}$$
$$+ \frac{\rho\alpha_0 Q_\psi^2}{h^2\psi^3\Delta\theta^2} + 3\rho\beta\omega_1^2\psi\left(1 - \frac{\cos^2\beta_0}{\sin^2\beta_0}\right)$$
$$- \frac{2\rho\delta Q_\psi\omega_1\sin\beta_0}{h\psi\Delta\theta\cos\beta_0} + \frac{\rho\gamma Q_\psi\omega_1\cos\beta_0}{h\psi\Delta\theta\sin\beta_0} - \frac{12\eta Q_\psi}{h^3\Delta\theta}$$
$$+ \frac{6\eta\omega_1\psi^2\cos\beta_0}{h^2\sin\beta_0}, \tag{19}$$

where$\varepsilon = 2/15$ [1, 2, 4].

Integrating the differential equation (19), for both regions of the SATBESPIG, where $h=h_1$ = const. and $h=h_2$ = const. (Figures 4 and 5), we obtain the calculus relations for the pressures ph_1 and ph_2:

$$
\begin{aligned}
p_{h1} \cong p_{\text{supp.}} &+ \frac{\rho\alpha_0 Q_\psi^2}{2h_1^2\Delta\theta^2}\left[\frac{1}{(L+\psi_0)^2} - \frac{1}{(L_2+\psi_0)^2}\right] \\
&+ \frac{3}{2}\rho\beta\omega_1^2\left(1-\frac{\cos^2\beta_0}{\sin^2\beta_0}\right)\left[(L_2+\psi_0)^2-(L+\psi_0)^2\right] \\
&- \frac{2\rho\delta Q_\psi\omega_1\sin\beta_0}{h_1\Delta\theta\cos\beta_0}\ln\frac{L_2+\psi_0}{L+\psi_0} \\
&+ \frac{\rho\gamma Q_\psi\omega_1\cos\beta_0}{h_1\Delta\theta\sin\beta_0}\ln\frac{L_2+\psi_0}{L+\psi_0} \\
&+ \frac{12\eta Q_\psi}{h_1^3\Delta\theta}\left[(L+\psi_0)-(L_2+\psi_0)\right] \\
&- \frac{2\eta\omega_1\cos\beta_0}{h_1^2\sin\beta_0}\cdot\left[(L+\psi_0)^3-(L_2+\psi_0)^3\right],
\end{aligned}
\tag{20}
$$

$$
\begin{aligned}
p_{h2} \cong p_{\text{supp.}} &- \frac{\rho\alpha_0 Q_\psi^2}{2h_2^2\Delta\theta^2}\left[\frac{1}{(L_2+\psi_0)^2} - \frac{1}{\psi_0^2}\right] \\
&+ \frac{3}{2}\rho\beta\omega_1^2\left(1-\frac{\cos^2\beta_0}{\sin^2\beta_0}\right)\left[(L_2+\psi_0)^2-\psi_0^2\right] \\
&+ \frac{2\rho\delta Q_\psi\omega_1\sin\beta_0}{h_2\Delta\theta\cos\beta_0}\ln\frac{\psi_0}{L_2+\psi_0} \\
&- \frac{\rho\gamma Q_\psi\omega_1\cos\beta_0}{h_2\Delta\theta\sin\beta_0}\ln\frac{\psi_0}{L_2+\psi_0} \\
&+ \frac{12\eta Q_\psi}{h_2^3\Delta\theta}\left[\psi_0-(L_2+\psi_0)\right] \\
&+ \frac{2\eta\omega_1\cos\beta_0}{h_2^2\sin\beta_0}\left[(L_2+\psi_0)^3-\psi_0^{\,3}\right].
\end{aligned}
$$

These relations are valid for the SATBESPIG with inner pumping. In a similar way, for the SATBESPIG with exterior pumping, we obtain

$$p_{h1} \cong p_{\text{supp.}} + \frac{\rho\alpha_0 Q_\psi^2}{2h_1^2\Delta\theta^2}\left[\frac{1}{\psi_0^2} - \frac{1}{(L_1+\psi_0)^2}\right]$$
$$+ \frac{3}{2}\rho\beta\omega_1^2\left(1 - \frac{\cos^2\beta_0}{\sin^2\beta_0}\right)\left[(L_1+\psi_0)^2 - \psi_0^2\right]$$
$$+ \frac{2\rho\delta Q_\psi\omega_1\sin\beta_0}{h_1\Delta\theta\cos\beta_0}\ln\frac{\psi_0}{L_1+\psi_0}$$
$$- \frac{\rho\gamma Q_\psi\omega_1\cos\beta_0}{h_1\Delta\theta\sin\beta_0}\ln\frac{\psi_0}{L_1+\psi_0} \tag{21}$$
$$+ \frac{12\eta Q_\psi}{h_1^3\Delta\theta}\left[\psi_0 - (L_1+\psi_0)\right]$$
$$- \frac{2\eta\omega_1\cos\beta_0}{h_1^2\sin\beta_0}\left[\psi_0^3 - (L_1+\psi_0)^3\right],$$

$$p_{h2} \cong p_{\text{supp.}} + \frac{\rho\alpha_0 Q_\psi^2}{2h_2^2\Delta\theta^2}\left[\frac{1}{(L+\psi_0)^2} - \frac{1}{(L_1+\psi_0)^2}\right]$$
$$+ \frac{3}{2}\rho\beta\omega_1^2\left(1 - \frac{\cos^2\beta_0}{\sin^2\beta_0}\right)\left[(L_1+\psi_0)^2 - (L+\psi_0)^2\right]$$
$$- \frac{2\rho\delta Q_\psi\omega_1\sin\beta_0}{h_2\Delta\theta\cos\beta_0}\ln\frac{L_1+\psi_0}{L+\psi_0}$$
$$+ \frac{\rho\gamma Q_\psi\omega_1\cos\beta_0}{h_2\Delta\theta\sin\beta_0}\ln\frac{L_1+\psi_0}{L+\psi_0} \tag{22}$$
$$+ \frac{12\eta Q_\psi}{h_2^3\Delta\theta}\left[(L+\psi_0) - (L_1+\psi_0)\right]$$
$$+ \frac{2\eta\omega_1\cos\beta_0}{h_2^2\sin\beta_0}\left[(L_1+\psi_0)^3 - (L+\psi_0)^3\right].$$

In relations (20)–(22), all the variables are known except the volumetric flow rate $Q\psi$.

CALCULUS RELATION FOR THE FLUID VOLUMETRIC FLOW RATE Q_ψ

The calculus relation for the fluid volumetric flow rate$Q_\psi$will be established using differential equation (19) again. To find the calculus relation, at the position of the radial step of thebearing we suppose that $dh/d\psi \neq 0$. So, in this zone of thebearing the inertial effects are dominant in comparison to theviscous effects. In other words, in the radial step zone of the bearing, the liquid moves approximately like an ideal but not viscous fluid [3, 5].

Integrating (19) in the vicinity of the radial step of the bearing grooved surface (Figures 4 and 5), we obtain a relation between ph_1, ph_2, and Q_ψ:

$$\Delta p = p_{h1} - p_{h2} \cong \frac{\rho Q_\psi^2}{\psi_c^2 \Delta\theta^2}(\alpha_0 + \varepsilon\psi_c)\left(\frac{1}{2h_2^2} - \frac{1}{2h_1^2}\right) - \rho\beta\omega_1^2\psi_c^2 \frac{\cos^2\beta_0}{\sin^2\beta_0}\ln\frac{h_1}{h_2}, \tag{23}$$

whereψc is the length measured by the ψ coordinate corresponding to the grooves radius of the profiled surface, rc (Figures 4 and 5). Relation (23) has some limits, especially at high values of h_1/h_2, when$h_2 \to 0$, or, in other words, at the heavy regimesfor the bearing functionality. If we know the other variables, including the extreme pressures ph_1 and ph_2, relation (23) offers the flow rate Q_ψ.

Therefore, using (20) and then (21), (22), and (23), we obtain a typical second degree algebraic equation, (24), from which we obtain the fluid flow rate Q_ψ:

$$
\begin{aligned}
&\frac{\rho Q_\psi^2}{\Delta\theta^2}\left\{\alpha_0\left[\frac{1}{2h_1^2}\left(\frac{1}{(L+\psi_0)^2}-\frac{1}{(L_2+\psi_0)^2}\right)\right.\right.\\
&\qquad\left.+\frac{1}{2h_2^2}\left(\frac{1}{(L_2+\psi_0)^2}-\frac{1}{\psi_0^2}\right)\right]\\
&\qquad\left.-\frac{1}{\psi_c^2}(\alpha_0+\varepsilon\psi_c)\left(\frac{1}{2h_2^2}-\frac{1}{2h_1^2}\right)\right\}\\
&+\frac{Q_\psi}{\Delta\theta}\cdot\left[\frac{\rho\gamma\omega_1\cos\beta_0}{\sin\beta_0}\left(\frac{1}{h_1}\ln\frac{L_2+\psi_0}{L+\psi_0}+\frac{1}{h_2}\ln\frac{\psi_0}{L_2+\psi_0}\right)\right.\\
&\qquad-\frac{2\rho\delta\omega_1\sin\beta_0}{\cos\beta_0}\left(\frac{1}{h_1}\ln\frac{L_2+\psi_0}{L+\psi_0}+\frac{1}{h_2}\ln\frac{\psi_0}{L_2+\psi_0}\right)\\
&\qquad\left.+12\eta\left(\frac{L-L_2}{h_1^3}+\frac{L_2}{h_2^3}\right)\right]-T1+T2\cong 0,
\end{aligned}
\tag{24}
$$

Where

$$
\begin{aligned}
T1 &= \frac{3}{2}\rho\beta\omega_1^2\left(1-\frac{\cos^2\beta_0}{\sin^2\beta_0}\right)\left[(L+\psi_0)^2-\psi_0^2\right],\\
T2 &= \rho\beta\omega_1^2\psi_c^2\frac{\cos^2\beta_0}{\sin^2\beta_0}\ln\frac{h_1}{h_2}-\frac{2\eta\omega_1\cos\beta_0}{\sin\beta_0}\\
&\quad\times\left\{\frac{1}{h_1^2}\left[(L+\psi_0)^3-(L_2+\psi_0)^3\right]\right.\\
&\quad\left.+\frac{1}{h_2^2}\left[(L_2+\psi_0)^3-\psi_0{}^3\right]\right\}.
\end{aligned}
\tag{25}
$$

Relation (24) is valid for the SATBESPIG with inner pumping. For the SATBESPIG with exterior pumping, we obtain a similar relation:

$$\frac{\rho Q_\psi^2}{\Delta\theta^2}\left\{\alpha_0\left[\frac{1}{2h_2^2}\left(\frac{1}{(L+\psi_0)^2}-\frac{1}{(L_1+\psi_0)^2}\right)\right.\right. \tag{26}$$

$$\left.-\frac{1}{2h_1^2}\left(\frac{1}{\psi_0^2}-\frac{1}{(L_1+\psi_0)^2}\right)\right] \tag{27}$$

$$\left.+\frac{1}{2\psi_c^2}(\alpha_0+\varepsilon\psi_c)\left(\frac{1}{h_2^2}-\frac{1}{h_1^2}\right)\right\}$$

Both (24) and (28) are classical algebraic equations of second degree in Q_ψ. If we denote by Q_ψ^I and Q_ψ^{II} the two solutions of the every algebraic equation (24) or (28), using Q_ψ^I and Q_ψ^{II}, and if we take into consideration that theSATBESPIG realizes inner pumping and exterior pumping, it is not possible to have a negative fluid flow rate from a physical point of view. Thus, the algebraic solution, which has physical meaning, is the positive solution [3, 5]. The numerical evaluation of Q_ψ^I and Q_ψ^{II} algebraic solution , for Q_ψ^I and Q_ψ^{II} using medium (normal) values for the physical and geometrical dimensions [1, 2] of (24) and (28), leads to $Q_\psi^I < 0$ and $Q_\psi^{II} > 0$. Thus, the mathematical relation for the calculus of the fluid volumetric flow rate Q_ψ is

$$Q_\psi \equiv Q_\psi^{II}, \tag{28}$$

where, in conformity to the devoted notations from theclassical algebra, the solution Q_ψ^{II} is

$$Q_\psi^{II} = \frac{-\ddot{b} - \sqrt{\ddot{b}^2 - 4\ddot{a}\ddot{c}}}{2\ddot{a}}, \tag{29}$$

Where$\ddot{a}$,b, and$\ddot{c}$are the coefficients of the algebraic equations (24) or (26).

NUMERICAL AND EXPERIMENTAL RESULTS

The established mathematical relations allow the numerical calculation of the pressure and speed distributions for several SATBESPIG with inner or external pumping in permanent and laminar regimes. For these numerical calculations weused two different computer programs: one for the SATBESPIG with inner pumping and the other for the SATBESPIG with external pumping. In Figures 6 to 8, we present the calculated pressure and speed distributions for a SATBESPIG with inner pumping and r_e = 90 mm, r_c = 57.15 mm, r_i = 45 mm, n_p = 10, α= 0.615, β_0 = 17°, ρ = 905 kg/m^3, η = 10^{-4}Pa·s, $p_{supp.}$ = 101325 N/m^2 , n = 900 rot/min, α_0 = 6/5, ε = 2/15, γ = 1/5, δ = 1/10, β = 2/15, and the parameters h1 and h2 as written in the figures.

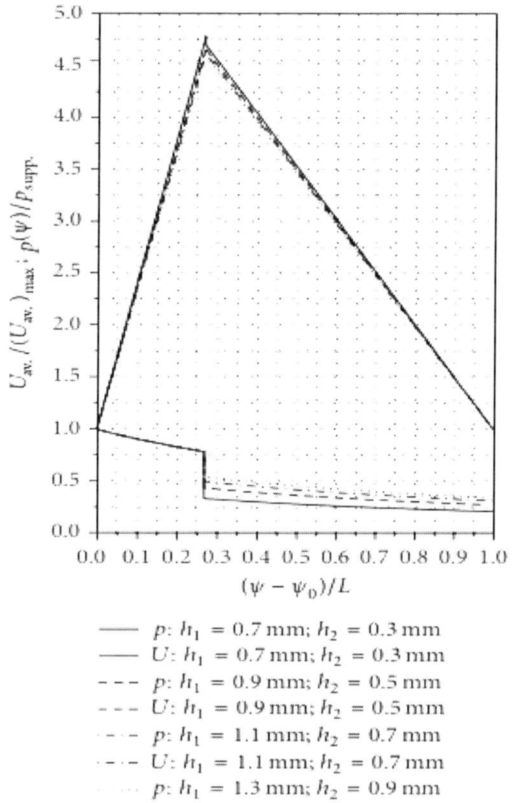

Figure 6: Pressure and speed distributions in the SATBESPIG with inner pumping. (See details in Figures 7 and 8.)

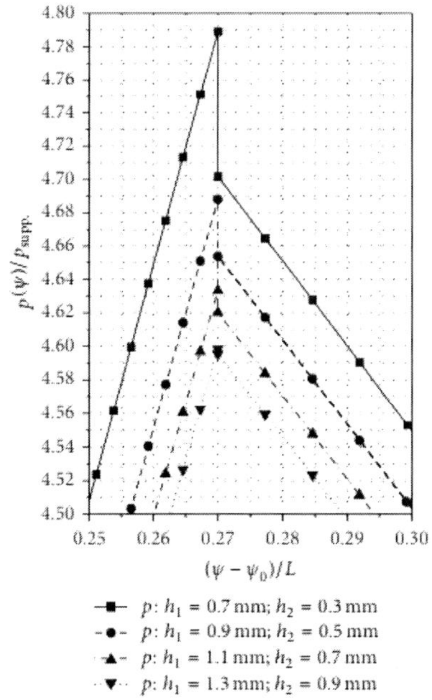

Figure 7: Detail of Figure 6 concerning the pressure jump between h_1 and h_2. (The marked points are those where we made the calculation.).

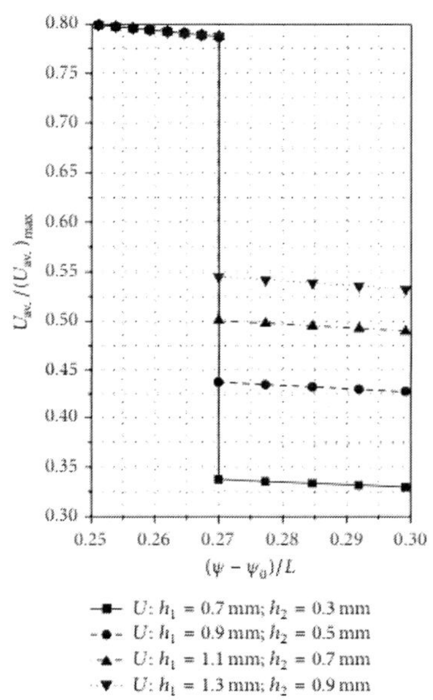

Figure 8: Detail of Figure 6 concerning the speed jump between h_1 and h_2. (The marked points are those where we made the calculation.).

Figure 9 presents the calculated pressure and speed distributions for a SATBESPIG with external pumping with the same characteristics as the SATBESPIG with inner pumping above, with the exception that r_c = 77.85 mm.

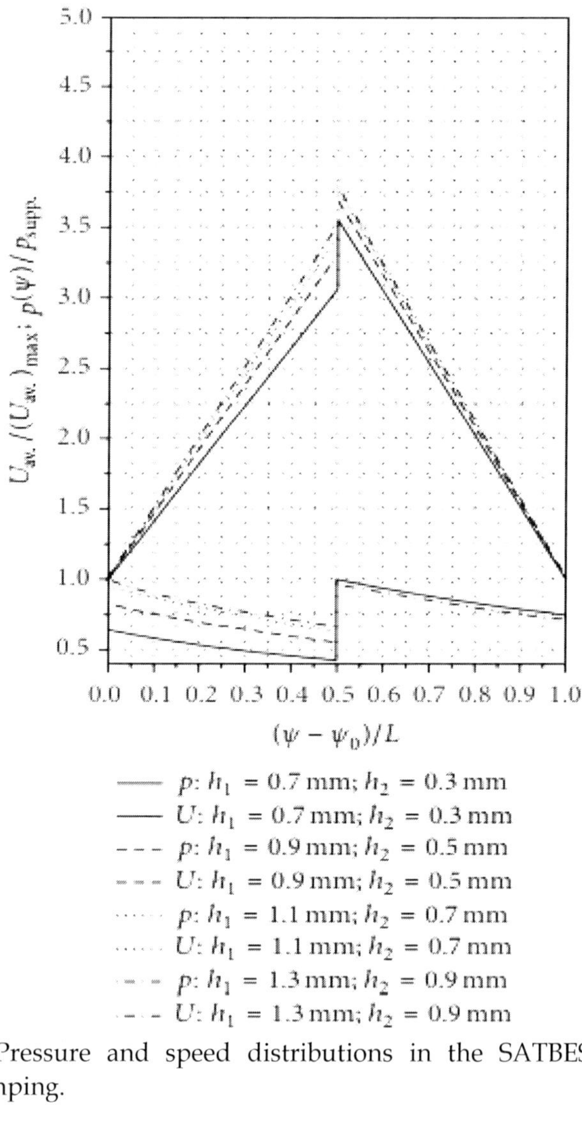

Figure 9: Pressure and speed distributions in the SATBESPIG with external pumping.

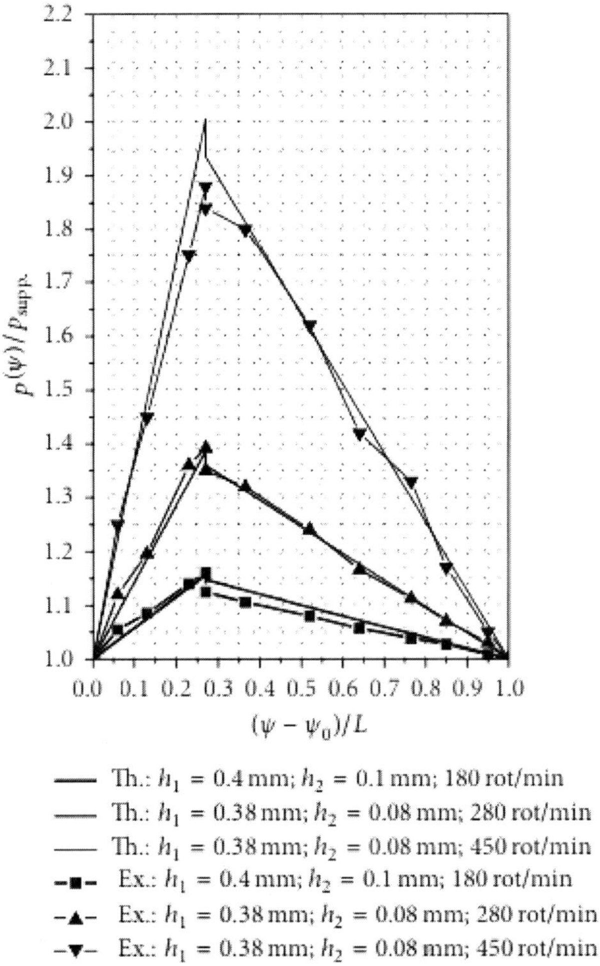

Figure 10: Pressure distributions in the SATBESPIG with inner pumping (comparison between the theoretical and experimental curves). The marked points are those where we performed the experimental measurements.

In Figure 10, we compared the theoretical results and the experimental measurements for the SATBESPIG with inner pumping and h_1, h_2, and $n$written in the figure. In all our studies, the main constructive geometrical and functional parameters for the calculation of these bearings were h_1, h_2, n, r_i, and r_e [3, 5].

DISCUSSION AND CONCLUSIONS

The analysis of the theoretical results allows some conclusionsto be drawn concerning the laminar motion of incompressible fluids in some models of a SATBESPIG. We analyzed two variants of the SATBESPIG with inner pumping, which are called the First Variant and the Second Variant below. The difference between these bearings is the r_c dimension.The First Variant is the SATBESPIG from Figure 6 (r_c = 57.15 mm), and the Second Variant has r_c = 70.29 mm.

(a) For all the rotation per minute (r.p.m.) ranges n and for all the analyzed rapports $h1/h2$, the First Variant SATBESPIG with inner pumping had superior hydrodynamics performance compared to the Second Variant SATBESPIG. Thus, high pressures in bearings can be realized with an appropriate dimension of the bearing surface, especially the dimension of the grooved surface, meaning the parameters r_i, r_c, and r_e.

(b) Generally speaking, increasing other hydraulics performance of the SATBESPIG for the same dimensions ri, rc, and re can be realized by increasing the number of the rotation per minute n. The modification of the rapport $h1/h2$ has a small influence.

(c) The non-dimensional speeds increase (and decrease) from the bearing input to the bearing output, with a jump at the radial step $r \cong rc$. The non-dimensional pressures change, but not linearly with the channel length, from the input, where $p_{imput} \cong p_{supp.}$, up to the pressure $ph1$ and from the pressure $ph2$ to the output pressure, where $poutput \cong psupp$..

(d) The pressure jump $\Delta p = ph_1 - ph_2$ theoretically tends to zero when the rapport $h_1/h_2 \rightarrow 1$ and thus when the radial step vanishes or when $ph_1 \cong ph_2$. The case $ph_1 \cong ph_2$ appears only when we do not take into consideration the inertial effects.

(e) The pressure jump Δp depends not only on the rapport h_1/h_2 but also on the value of the n and on the bearing geometry (ri, rc, and re).

(f) For the same constructive geometrical and functional parameters (r_i, r_e, h_1/h_2, n,...), the SATBESPIG with exterior pumping gives lower pressures than the bearing with inner pumping.

(g) The comparative analysis between the theoretical and experimental results shows good correlation, especially at low n

and at middle values for the rapport h_1/h_2 ($h_1/h_2 \cong 4$). Some of the mathematical relations established above can be used to approach the theoretical functionality of the SATBESPIG having magnetic controllable fluids (magnetic fluids or magnetorheological fluids) as incompressible fluids, in the presence of a controllable magnetic field [18- 20].

CONFLICT OF INTERESTS

The authors declare that there is no conflict of interests regarding the publication of this paper.

ACKNOWLEDGMENT

This work was partly supported by the Project PN II (National Education Ministry—Romania) no. 157/2012 "MagNanoMicroSeal" (2012–2015).

REFERENCES

1. C. Majumdar, R. Pai, and D. J. Hargreaves, "Analysis of water-lubricated journal bearings with multiple axial grooves," Proceedings of the Institution of Mechanical Engineers J, vol. 218, no. 2, pp. 135–146, 2004.
2. C. Velescu, "Theoretically remarks on the thermal and visco—inertial effects in the self acting sliding conically radially—thrust bearings," in Proceedings of the International Symposium on Fluid Machinery and Fluid Engineering (ISFMFE '96), Beijing, China, September 1996.
3. C. Velescu, Hydrodynamics of the Thrust Bearings With Spiral Grooves, Mirton Publishing House, Timisoara, Romania, 2000.
4. C. Velescu, The study of the viscous fluids motion in self-acting thrust bearings [Ph.D. thesis], "Politehnica" University of Timisoara, 1998.
5. De Sabata, N. C. Popa, I. Potencz, and L. Vékás, "Inductive transducers with magnetic fluids," Sensors and Actuators A, vol. 32, no. 1–3, pp. 678–681, 1992.

6. G. I. Broman, "Implications of cavitation in individual grooves of spiral groove bearings," Proceedings of the Institution of Mechanical Engineers J, vol. 215, no. 5, pp. 417–424, 2001.
7. J. F. Zhou, B. Q. Gu, and C. Ye, "An improved design of spiral groove mechanical seal," Chinese Journal of Chemical Engineering, vol. 15, no. 4, pp. 499–506, 2007.
8. L. Wu, P. Liu, Z. Liu, Z. Liu, Q. Yang, and Y. Wei, "Optimization design of hydrodynamic bearing with multiple spiral grooves," in Proceedings of the 7th International Conference on Progress Machining Technology (ICPMT '04), pp. 502–507, Suzhou, China, December 2004.
9. M. Liță, N. C. Popa, C. Velescu, and L. N. Vékás, "Investigations of a magnetorheological fluid damper," IEEE Transactions on Magnetics, vol. 40, no. 2, pp. 469–472, 2004.
10. M. Rodkiewicz and M. I. Anwar, "Inertia and convective effects in hydrodynamic lubrication of a slider bearing," Journal of Lubrication Technology F—ASME Transactions, vol. 93, 1971.
11. M. Rodkiewicz, J. C. Hinds, and C. Dayson, "Inertia, convection, and dissipation effects in the thermally boosted oil lubricated sliding thrust bearing," Journal of Lubrication Technology F—ASME Transactions, vol. 97, no. 1, pp. 121–129, 1975.
12. N. C. Popa, I. De Sabata, I. Anton, I. Potencz, and L. Vékás, "Magnetic fluids in aerodynamic measuring devices," Journal of Magnetism and Magnetic Materials, vol. 201, no. 1–3, pp. 385–390, 1999.
13. N. Kawabata, I. Ashino, M. Sekizawa, and S. Yamazaki, "Spiral grooved bearing utilizing the pumping effect of a herringbone journal bearing. (Method of numerical calculation and influences of bearing parameters)," JSME International Journal 3, vol. 34, no. 3, pp. 411–418, 1991.
14. S. B. Malanoski and C. H. T. Pan, "The static and dynamic characteristics of the spiral—grooved thrust bearing," Journal of Basic Engineering D—Transactions of the ASME D, vol. 87, no. 3, pp. 547–558, 1965.
15. S. M. Yao, "Aerostatic and aerodynamic performance of an in-pump spirally grooved thrust bearing: Analysis and comparisons to static load experiments," Transactions, vol. 51, no. 5, pp. 679–689, 2008.

16. S. M. Yao, "Aerostatic and aerodynamic performance of an out-pump spirally grooved thrust bearing: analysis and comparisons to static load experiments," Tribology Transactions, vol. 52, no. 3, pp. 376–388, 2009.
17. V. N. Constantinescu, Al. Nica, M. D. Pascovici, G. Ceptureanu, and S. Nedelcu, Sliding Bearings, Tehnica Publishing House, Bucharest, Romania, 1980.
18. V. N. Constantinescu, Dynamics of Viscous Fluids in Laminar Regime, Romanian Academy Publishing House, Bucharest, Romania, 1987.
19. V. Yemelyanov and I. A. Yemelyanov, "Physical models, theory and fundamental improvement to self-acting spiral-grooved gas bearings and visco-seals," Proceedings of the Institution of Mechanical Engineers J, vol. 213, no. 4, pp. 263–271, 1999.
20. W. C. Wachmann, M. S. Malanoski, and V. J. Vohr, "Thermal distortion of spiral-grooved gas-lubricated thrust bearings," Journal of Lubrication Technology—ASME Transactions, vol. 93, no. 1, pp. 102–112, 1971.

CITATION

Cornel Velescu and NicolaeCalinPopa, "Laminar Motion of the Incompressible Fluids in Self-Acting Thrust Bearings with Spiral Grooves," The Scientific World Journal, vol. 2014, Article ID 478401, 10 pages, 2014. doi:10.1155/2014/478401.

CHAPTER 3

Effects of Air SAC Compliances on Flow in the Parabronchi: Computational Fluid Dynamics Using an Anatomically Simplified Model of an Avian Respiratory System

Akira Urushikubo, Masanori Nakamura, Hiroyuki Hirahara

Department of Mechanical Engineering, Saitama University, Saitama, Japan

Keywords

Computational Fluid Dynamics; Avian Lung; Air Sacs; Compliance

ABSTRACT

Air flow in an avian lung was studied numerically to determine the effects of air sac compliance on flow in the parabronchi. In this preliminary study, the geometry of the avian respiratory system was simplified to capture the characteristics of respiratory flow. The pressure fluctuation within air sacs caused by inflation and deflation was expressed by a lumped parameter model. The results demonstrate that the flow direction in the parabronchi varied, depending upon the compliance of the air sacs. A unidirectional flow in the parabronchi was achieved for compliances where pressure fluctuations in all air sacs were in phase. Air sac compliance significantly affected the pressures in the anterior and posterior air sacs and thus the pressure difference over the parabronchi that drove the flow in the parabronchi. These results address the importance of air sac compliance in the avian respiratory system and suggest that the compliance of air sacs would be optimized to accomplish unidirectional flow in the parabronchi.

INTRODUCTION

Avian respiratory systems mainly consist of the trachea, bronchi and air sacs. The bronchi form a continuous loop, starting from the mesobronchus, ventrobronchi, parabronchi, and ending up with the dorsobronchi [1-3]. This loop structure allows inhaled air to flow in only one direction. Oxygen-rich inhaled air passes, during both inspiration and expiration, through the parabronchi, where an exchange of oxygen and carbon dioxide takes place [4-6]. This lung design helps birds maintain high metabolic rates for flight at altitude [6,7].

Two mechanical factors have been proposed to induce unidirectional flow in the parabronchi. One is an aerodynamics valve. As shown by radiophotography and microscopy [8,9], the primary bronchi just cranial to the ventrobronchial junction, the junction between primary bronchus and ventrobronchi, is narrowed in some birds. This narrowing accelerates the inhaled air, thereby increasing its convective inertia and directing the inhaled air to the mesobronchi [8,10-12]. It also causes flow separation at the ventrobronchial junction [12] and enhances formation of vortices at the origin of the ventrobronchi. The vortices work as a type of lid for the inhaled air, such that it cannot flow into the ventrobronchi [13-15]. The other mechanical factor is a pressure difference between the air sacs [5,16]. Evidence for this lies in the experiments [17,18] that demonstrate that occlusion of air sacs with cotton results in depression of the respiratory function, although the degree of depression depended in that work on the combination of occlusions. On the basis of these results, the pressure difference between air sacs was concluded to regulate the flow in parabonchi. The air sacs are stretchy bags composed of thin membranes [3,19]. They deform largely and repeatedly in accordance with a change in pleural pressure to store in and push out air for helping continuous flow in parabronchi where exchange of oxygen and carbon dioxide takes place [20].

Airsacculitis is one of the pathological symptoms caused by Escherichia coli [21], Mycoplasma [22-24], aspergillosis [25], and so forth. Infected by these bacteria and virus, the air sac tissue is histologically changed and gets thickened [22-24]. It is also known that birds with airsaccultis exhibit respiratory distress [26]. Considering these two facts together, we postulate that the compliance of air sacs would be optimized for the activities of each bird species, and the change in air sac compliance would result in changes in respiratory flow and accordingly respiratory distress.
The aim of the study was to investigate the effects of air sac compliance on flow in the parabronchi. In this preliminary study, the geometry of the avian respiratory system was simplified to capture the characteristics of respiratory flow. The pressure fluctuation within air sacs caused by their inflation and deflation was expressed with a lumped parameter model. The flow in the avian respiratory system was studied numerically using various values of air sac compliance to determine their influence on respiratory flow.

METHOD

Geometry of the Avian Lung Model

Figure 1 presents an overview of the model of the avian respiratory system. The model simplifies the anatomical characteristics of the avian lung [4]. Here we assume that the geometry and diameters of the trachea and bronchi do not change during the respiratory cycle for the following reasons. First, the trachea is supported by rings of cartilage [1,9,27] and airways embedded in parenchyma are practically rigid [3,4,28-30]. Second, the lung volume varies by a mere 1.4% between respiratory cycles [31]. The connecting points and length of the respiratory bronchi were determined on the basis of anatomical data [1,2, 32,33]. For simplicity, the diameter of the bronchi, D, was uniformly set to 4.5 mm in reference to the maximum diameter of the primary bronchi of the domestic fowl [1]. The cross-section identified by the arrow in Figure 1(a) represents the inlet connecting to the syrinx which in turn connects to the trachea and then the larynx. For further explanation, a Cartesian coordinate system (x, y, z) is defined at the origin, O, placed at the center of the inlet, such that the x axis lies along the center line of the primary bronchi. In the following, the nomenclature of [4] is used. Figure 1(b) presents the x-y cross-section of the model with a description of the lengths. The channel extending along the x-axis from the inlet is the primary bronchus, with a length of 24.55 mm. The primary bronchus bifurcates into the mesobronchus, which extends along the x-axis and the ventrobronchi, which follows the y-axis. The mesobronchus is further connected to the dorsobronchi, which is established along the y-axis. Two posterior air sacs, namely, the posterior thoracic air sac

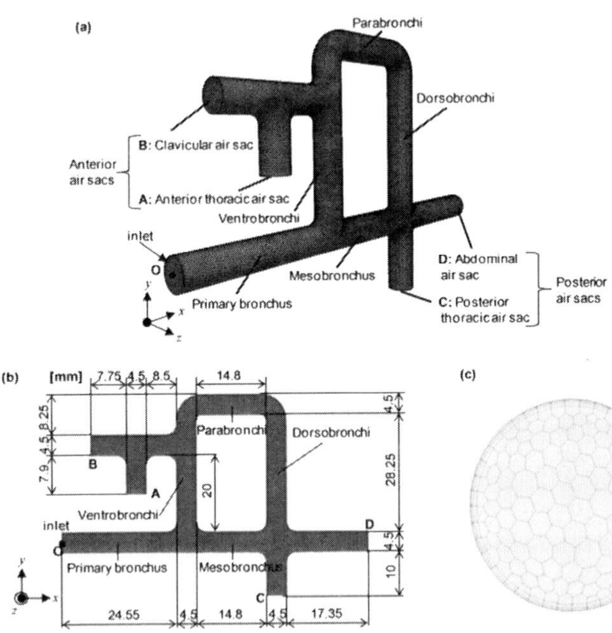

Figure 1: Anatomically-simplified model of the avian lung. (a) The computational mesh; (b) Cross-section view of the flow channel at the x-y plane; and (c) The computational mesh of the inlet cross-section.

(cross-section C) and the abdominal air sac (cross-section D), are present at the junction between the mesobronchus and the dorsobronchi. The dorsobronchi turns left to connect with the parabonchi, where the exchange of oxygen and carbon dioxide takes place. The parabrochi meets the ventrobronchi, which has two anterior air sacs, namely, the anterior thoracic air sac (cross-section A) and the clavicular air sac (cross-section B) that includes the interclavicular air sac and the cervical air sac.

Computational meshes were created for flow analysis (Figure 1(a)). Four layers of prism meshes that account for 7.6% in the radial direction from the wall were allocated near the wall to resolve boundary layers. The rest of the computational domain was filled with polyhedral meshes, as seen in Figure 1(c), which shows the mesh of the inlet. The total number of nodal points was 387,781 and that of the meshes was 132,004.

Modeling of the Mechanics of an Air Sac

Enlargement of the body cavity by inspiratory muscle action lowers pressure in the air sacs relative to the atmosphere during inspiration, thereby drawing air into the lung and air sacs [3]. Once the elastic force of the air sac exceeds the pressure in the air sac, it deflates, thereby causing an outflow of the air, with a decrease in the pressure within the air sac. To represent this behavior, we employed a lumped parameter model, given by

$$p^k = \frac{\Delta V^k}{C^k} \tag{1}$$

where p^k is the pressure in air sac k, ΔV^k is the total change in the volume of air sac k, and C^k is the compliance of the air sac k. Superscript k stands for one of the air sacs attached to cross-section A, B, C or D. The total change in the volume of each air sac from its natural state was calculated from

$$\Delta V^k = \int_0^t q^k(T)\,\mathrm{d}T \tag{2}$$

where q(t) is the flow rate entering or exiting the air sac at time t. In Equation (1), the numerator stands for the total volume change from the natural state. This equation describes the situation where an increase in the total amount of air contained in the air sac causes the elevation of pressure within the air sac, and vice versa.

Flow Analysis

As described below, the maximum velocity at the inlet was set at 2.06 m/s. Using the speed of sound at room temperature (a = 343.51 m/s), the Mach number for the system was estimated to be 0.006. This is much smaller than the threshold, 0.3, below which it can be assumed that air is an incompressible fluid, with density 1.2 kg/m^3 and kinematic viscosity 1.5×10^{-5} m^2/s. Furthermore, flow was assumed to be laminar because the maximum Reynolds number assessed at the inlet was approximately 725. The flow simulations were performed by solving the Navier-Stokes equations and the equation of continuity, given by

$$\frac{DU}{Dt} = -\frac{1}{\rho}\nabla p + \nu\nabla^2 U \qquad (3)$$

$$\nabla \cdot U = 0 \qquad (4)$$

where U is the velocity vector (u, v, w) in the (x, y, z) directions and is a differential operator. The equations were solved with Open FOAM software (SGI Corp, v2.0.1) that adopts a finite volume method on the basis of the pressure implicit with splitting of operators method for formulation. An implicit Euler scheme was used for the unsteady term, the first-order Gaussian integration with a central difference scheme was used for the gradient term, and the first-order Gaussian integration with an upwind scheme was used for the divergence term [34]. Convergence of the numerical solution was assessed based on a maximum tolerance of 1.0×10^{-6} for pressure and $1.0 \times 10^{-5}$for velocity.

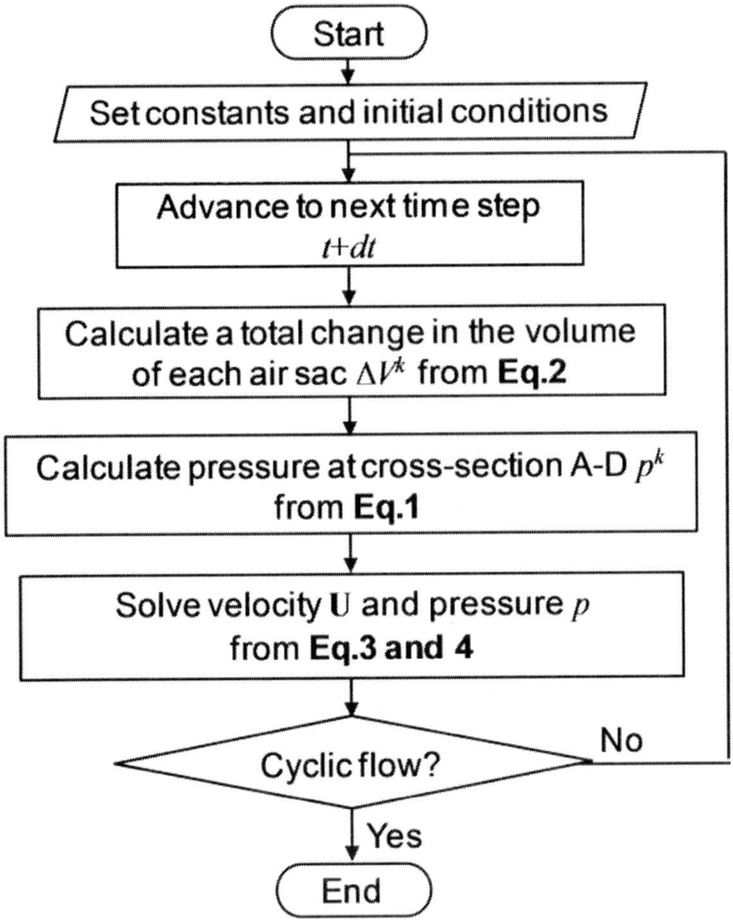

Figure 2: A flowchart of the avian respiratory flow simulation.

Simulation Procedure

A flowchart of the simulation is shown in Figure 2. The simulation was commenced from a quiescent flow state. With advancement of a time step, a total change in the volume of each air sac from its natural state was calculated from velocity obtained by the current time step from Equation (2), and then the pressure in each air sac was calculated from Equation (1). The pressure in each air sac was assigned to the cross-sections A-D as outlet boundary conditions for the following flow simulation. Inflow velocity was given at the inlet as an inlet boundary condition. Since effects of the air sac compliance on respiratory flow would be more manifested than eupneic breathing, panting breath condition was employed. Figure 3 plots a time variation of the inflow rate used in this study. The curve was obtained by approximating the data of panting breath of a duck with a sinusoidal function [20], and the frequency of respiration f was set to 10 Hz in reference to the data of domestic pigeons in panting breath [35]. The Womersley number assessed with the inlet diameter as a characteristic length α_w was 4.6. Because the flow was oscillatory, the velocity profile at the inlet was defined by Womersley flow [36]:

$$u = \mathbf{Re}\left(\frac{A}{i\rho}\frac{1}{2\pi f}\left\{1-\frac{J_0\left(\alpha_w \eta i^{3/2}\right)}{J_0\left(\alpha_w i^{3/2}\right)}\right\}e^{i2\pi ft}\right) \tag{5}$$

$$v = w = 0 \tag{6}$$

where Re means "take the real part of", f is a frequency of respiration, α_w is the Womersley number, J_0 is the Bessel function of the first kind, and η is the non-dimensional number r/R representing a normalized radial position with respect to the radius of the inlet R. Constant A is determined from the tidal volume V_T. The constant A in Equation (5) was determined as 211.8 kg/m^2s^2 from the tidal volume of a pigeon $V_T = 1.05 \times 10^{-6}$ m^3 [35]. The non-slip condition (zero velocity) was assumed at the wall. Given these boundary conditions, the flow simulation in the anatomically-simplified model of the avian lung was implemented by solving Equations (3) and (4). This process was repeated until cyclically repeatable flow was obtained. In actual simulations, the period of a respiratory cycle was divided into 200 steps with a time increment of $\Delta t = 5 \times 10^{-4}$ s for C equal to and larger than 1, and 1000 steps with a time increment of $\Delta t = 1 \times 10^{-4}$ s for the rest of conditions. According to the results, a cyclically repeatable flow was attained at most by the 13th cycle.

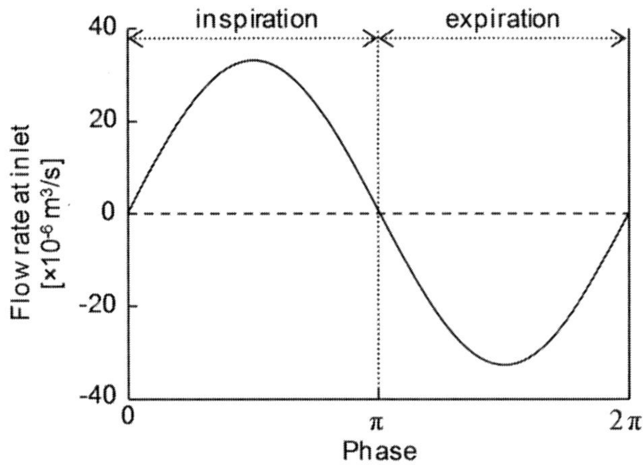

Figure 3: Time variation of the flow rate at the inlet.

We performed flow simulations with different values of the compliance. A unit value of the compliance C_0 was 8.158 × 10^{-8} m^3/Pa in reference to the compliance of the domestic fowl [37]. Note that this referred value of compliance is the overall compliance of the air sacs and the surrounding body walls. However, to our best knowledge, no published data on the compliance of air sacs is available. The values of the compliance of air sacs were changed concurrently, i.e. $C^A = C^B = C^C = C^D = C$, in 11 multiple of C_0 (0.01, 0.1, 0.25, 0.5, 1, 2, 5, 8, 10, 50, 100) to determine their effects on respiratory flow in the avian lung. Hereinafter, each condition was described with its multiple number of C_0, e.g. $C = 0.01C_0$ is $C = 0.01$.

RESULTS

Flow Pattern

Figure 4 illustrates streamlines in the avian lung under the condition C = 0.01 at t/T = (a) 1/8 (acceleration in inspiration), (b) 3/8 (deceleration in inspiration), (c) 5/8 (acceleration in expiration) and (d) 7/8 (deceleration in expiration). As shown in Figure 4(a), the air coming through the primary bronchus split into two

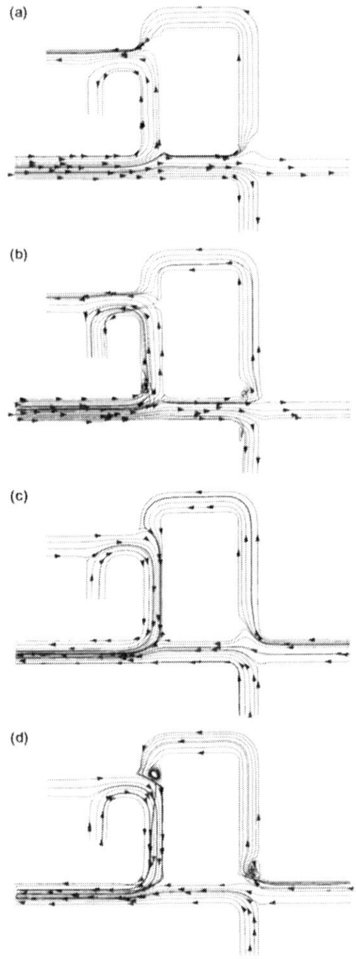

Figure 4: Streamlines in the avian lung at C = 0.01. (a) t/T = 1/8, (b) t/T = 3/8, (c) t/T = 5/8, (d) t/T = 7/8.

directions, the mesobronchus and the ventrobronchi. The flow heading to the mesobronchus was directed to the parabronchi from the posterior side to the anterior side (from the right to the left in Figure 4) after passing through the posterior air sacs and the dorsobronchi. In contrast, the flow heading to the ventrobronchi merged with the flow from the parabronchi and flowed into the anterior air sacs. The overall behavior of the flow during the deceleration of inspiration (**Figure 4**(b)) appeared almost the same as that observed during acceleration. However, vortices were formed at the corner of the primary bronchi and the ventrobronchi, as well as near the junction of the primary bronchi and the dorsobronchi.

When the flow turned to expiration, these vortices vanished, as seen in **Figure 4**(c). In this phase, part of the air ejected from the posterior air sacs was delivered to the parabronchi through the dorsobronchi, while the rest was directed to the mesobronchi. At the same time, the air ejected from the anterior air sacs merged with that from the parabronchi and later joined air from the mesobronchi to exit. Macroscopically, flow during the decaleration phase of expiration showed a similar pattern to that during acceleration, except that vortices formed at the corner of the dorsobronchi and the mesobronchi, and at the corner at the ventrobronchi and the parabronchi, as seen in Figure 4(d). At C = 0.01, the air in the parabrochi flowed from the anterior side to the posterior side throughout the respiratory cycle.

The flow pattern at C = 1 is shown in Figure 5. Time instants (a)-(d) in Figure 5 are the same as those in Figure 4. Although the overall behavior of the flow appears similar to that observed for C = 0.01, there are differences between Figures 4 and 5 worth noting. As in Figure 5(a), the flow in the parabronchi goes from the left to the right during acceleration in inspiration. The flow direction during deceleration was reversed from that during acceleration, as observed in Figure 5(b). At the same time, vortices formed at various corners. In particular, a large vortex was generated at the corner of the mesobronchi and the dorsobronchi. Those vortices were not observed during acceleration of expiration, as in Figure 5(c). During the deceleration phase of expiration, a small vortex was found at the root of the dorsobronchi (Figure 5(d)). In this case, flow in the parabronchi was not unidirectional; it was directed from the anterior to the posterior side during the acceleration of inspiration and from the posterior to the anterior side during the remainder of the respiratory cycle.

Figure 6 shows the flow pattern at C = 10. Again, time instants (a)-(d) in Figure 6 are the same as those in

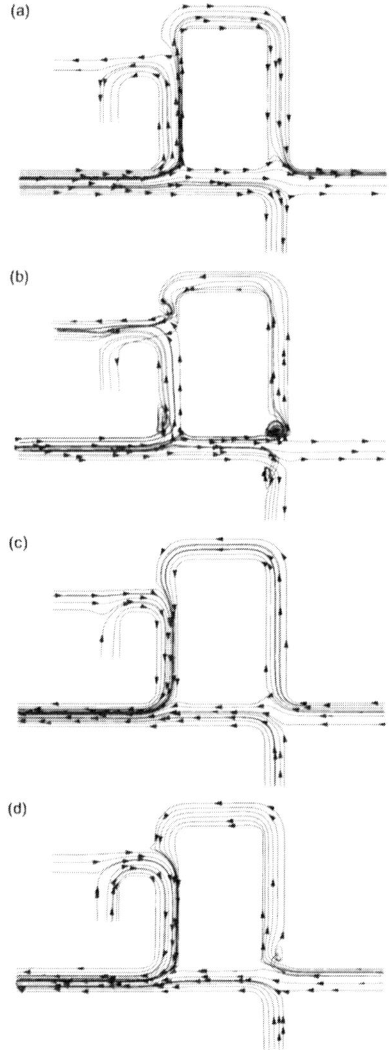

Figure 5: Streamlines in the avian lung at C = 1. (a) t/T = 1/8, (b) t/T = 3/8, (c) t/T = 5/8, (d) t/T = 7/8.

Figure 4, and the flow pattern at C = 10 was macroscopically the same as that at C = 0.01. Under this condition, the air in the parabonchi flowed from the posterior to the anterior side throughout the respiratory cycle. Differences between Figures 4 and 6 were found during the deceleration phase of both inspiration and expiration. During the deceleration phase of inspiration, the vortex that formed at the corner of the primary bronchi and the dorsobronchi occupied a larger space of the dorsobronchi, blocking the upwardly directed airflow (Figure 6(b)). During the deceleration phase of expiration, no vortices were observed (Figure 6(d)).

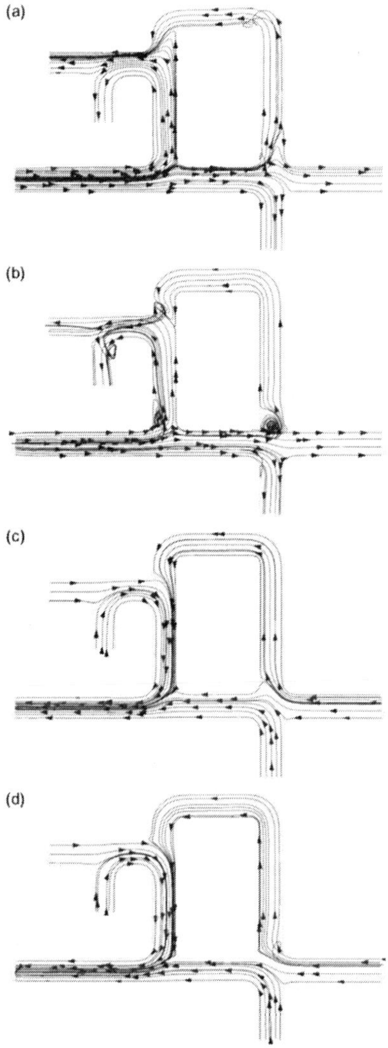

Figure 6: Streamlines in the avian lung at C = 10. (a) t/T = 1/8, (b) t/T = 3/8, (c) t/T = 5/8, (d) t/T = 7/8.

Flow in the Parabronchi

The time variation of the flow rate at the cross-section (x = 36.45 mm) of the parabronchi is shown in Figure 7. The horizontal axis represents the phase in the respiratory cycle. A positive flow rate value is defined for the air flowing from the posterior to the anterior side (the negative direction on the x-axis). For C = 1, the flow rate was negative during part of the inspiration phase and positive during the rest of the cycle. This means that the flow in the parabronchi was bidirectional for C = 1. Bidirectional flow was observed for C = 0.5 and 100 as well, although there was almost no negative flow for C = 100. In contrast, the flow in the parabronchi was always positive and thus unidirectional for C = 0.01 and 10. It was also found that the phase when the flow rate reached a maximum shifted to the expiration phase as the compliance of the air sacs increased.

Effects of Air Sac Compliance on the Generation of Unidirectional Flow in the Parabronchi

The total volumes of the positive flow V_+ and the negative flow V_- passing through the parabronchi over one respiratory cycle were calculated. The ratio of V_+ to V_- is plotted against the compliance of the air sacs in Figure 8. As V_-/V_+ approaches zero, the unidirectional nature of flow in the parabronchi becomes stronger. As in Figure 8, V_-/V_+ was zero for the low compliances C = 0.01 and 0.1 and for the high compliances C = 8, 10 and 50. The value of V_-/V_+ was the largest for C = 0.5, and decreased as the compliance became either smaller or larger. For C = 5 and 100, V_-/V_+ was not zero but was nevertheless quite small compared with that at C = 1.

Time Variation of a Pressure in Each Air Sac

Pressures at the cross-section A-D or the air sacs are plotted as a function of time in **Figure 9**. Figures 9 (a)- (e) represent the data obtained for compliances of 0.01, 0.5, 1, 10 and 100, respectively. As seen in Figure 9, the range of the pressure in the air sacs decreased

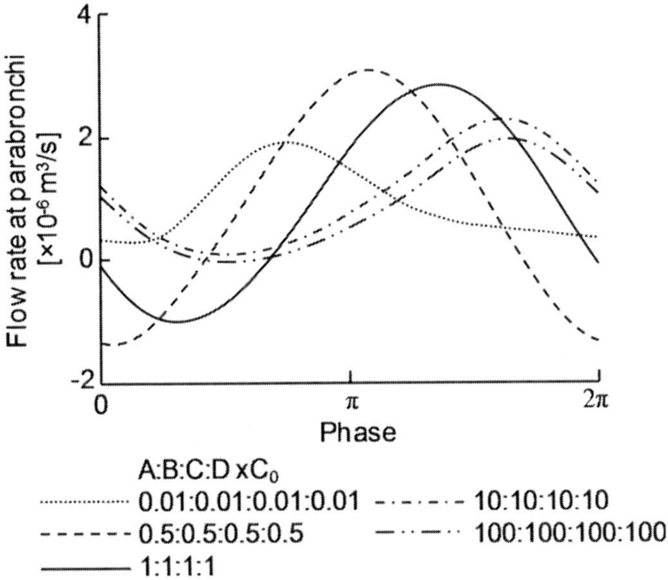

Figure 7. Temporal variation of the flow in the parabronchi.

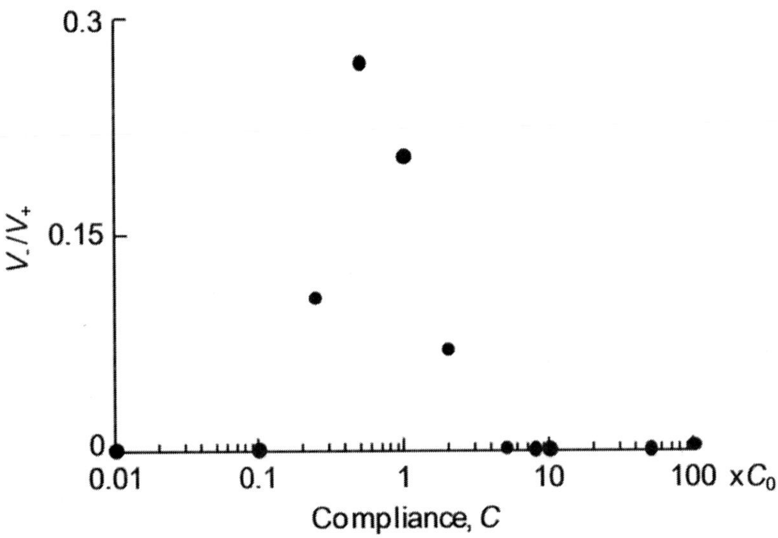

Figure 8: Plot of the ratio V-/V+ against air sac compliance.

as their compliances decreased. For C = 0.01, the curves appear to virtually overlap (Figure 9(a)). However, magnified views of the pressure curves revealed that the pressures in the posterior air sacs (C and D) were larger than those in the anterior airs sacs (A and B) during inspiration. During expiration, the pressure relation inverted, demonstrating larger pressure in the anterior air sacs. Figures 9(b) and (c) show a slightly different trend from Figure 9(a). At C = 1, the pressures in the posterior air sacs were larger than those in the anterior airs sacs during the deceleration phase of inspiration and the acceleration phase of expiration. For larger compliances (Figures 9(d) and (e)), the pressures in the abdominal air sac, D, were always larger than those in the anterior air sacs A and B throughout the respiratory cycle. For C = (a) 0.01, (d) 10 and (e) 100, the pressure in the air sacs became maximal at the phase π, corresponding to the turning point from inspiration to expiration.

Time Variation of a Pressure Difference over the Parabronchi

The pressure at the cross-section of the parabronchi at x = 31.4 mm and that at x = 42.0 mm were spatially averaged for each time and defined as P_{ante} and P_{post}, respectively. The time variations of the pressure difference $P_{post} - P_{ante}$ for C = 0.01, 0.5, 1, 10 and 100 are plotted inFigure 10. For C = 0.5 and 1, the absolute value of the negative maximum pressure difference was almost the same as that of the positive maximum pressure difference. In addition, the length of time during which the pressure difference was negative was equivalent to that for the positive pressure difference. In contrast, for C = 0.01 and 10, the absolute values of the negative maximum pressure differences were smaller than those of the positive maximum pressure differences, and the length of time for the negative pressure difference was shorter than that for the positive pressure difference.

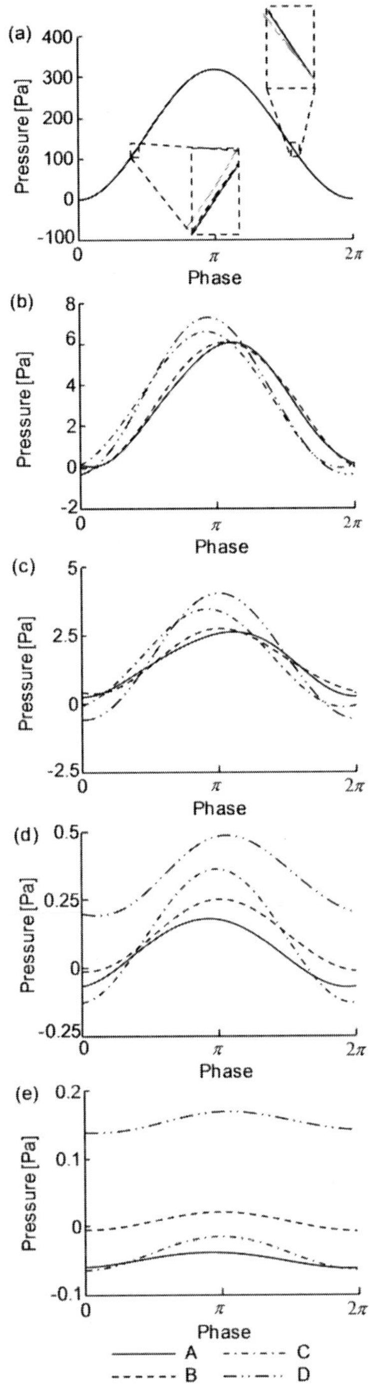

Figure 9: Time variation of the pressure in air sacs A-D. (a) C = 0.01, (b) C = 0.5, (c) C = 1, (d) C = 10, (e) C = 1009.

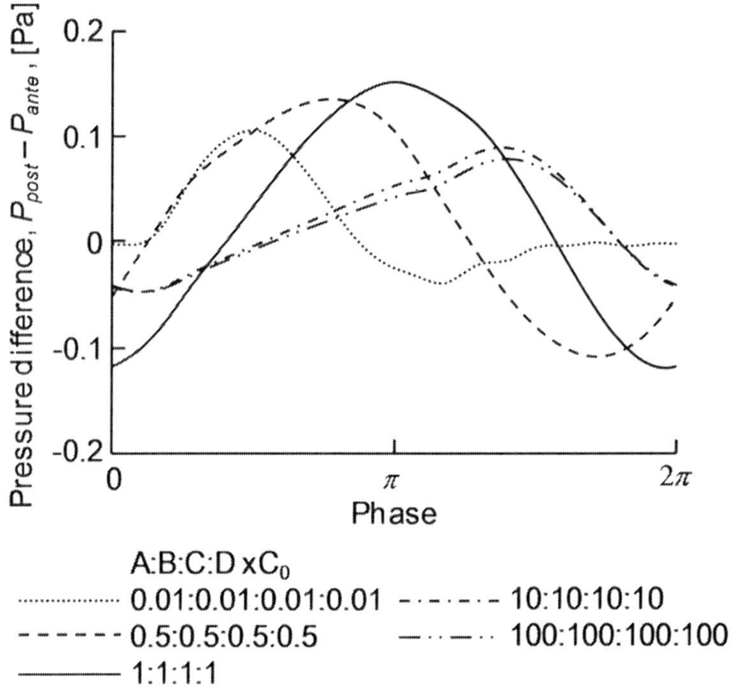

Figure 10: Time variation of the pressure difference over the parabronchi.

DISCUSSION

In the present study, an avian respiratory flow was simulated using an anatomically simplified model of avian lung. The obtained flow patterns in Figures 4-6 demonstrated that during inspiration, the main stream of inflow was lead to the posterior air sacs. During expiration, the air stocked in the anterior air sacs was directed to the primary bronchi while air in the posterior air sacs went to the parabronchi. In some conditions, flow in the parabronchi was unidirectional throughout a respiratory cycle. Such features of the main flow are consistent with the ones in a real avian lung [3-5,20,38,39]. It is generally considered that in real birds air barely flows in the ventrobronchi during inspiration and in mesobronchi during expiration [5, 8, 10, 11, 20, 38, and 39]. In contrast, here we observed not a little amount of flow in the ventrobronchi during inspiration and in mesobronchi during expiration. This is attributable to smaller contributions of aerodynamic valving caused by vortices and flow separations due to simplification of the geometry of the avian lung and the absence of segmentum accelerans as discussed in the following paragraphs. Although a quantitative comparison with previous studies to validate the results is difficult due to differences in various conditions including geometry, these results indicated that the model represented main characteristics of avian respiratory flow such as unidirectional flow in the parabronchi.

As shown in Figures 4-6, the flow direction in the parabronchi varied, depending upon the compliance of the air sacs. Figure 8 demonstrates that the achievement of unidirectional flow in the parabronchi was bimodal for small compliances (C = 0.01 and 0.1) and large compliances (C = 8, 10 and 50). By examining at Figure 9, we found that for those cases, the pressure fluctuations in all air sacs were in phase, whereas they were out of phase for the conditions under which the flow in the parabronchi was not unidirectional. According to Cohn and Shannon [40], such synchronization of the pressure in phase is a feature of the real avian lung. These data suggest that phase-synchronization of the pressure fluctuations in air sacs would be one of the necessary conditions to accomplish unidirectional flow in the parabronchi.

Figures 7, 8 and 10 demonstrate that the trend in pressure difference over the parabronchi differed between unidirectional (C = 0.01, 0.1, 8, 10 and 50) and bidirectional (C = 0.25, 0.5, 1, 2, 5 and 100) flow in the parabronchi. For the former case, the absolute value of the negative maximum pressure difference was smaller than that of the positive maximum pressure difference. In contrast, for the latter case, the absolute value of the positive maximum pressure difference was almost equal to that of the negative maximum pressure difference, and the time for the positive pressure difference was as long as that for the negative pressure difference. On the basis of these results, the driving mechanism for air flow in the parabronchi may be explained as follows. As the positive pressure difference over the parabronchi increases, air flow is driven from the dorsobronchi to the ventrobronchi. Once created, the convective inertial force acts to maintain this flow, even if the pressure difference becomes zero or even negative. Therefore, as long the negative pressure difference over the parabronchi was of short duration or the magnitude of the negative pressure was small, unidirectional flow was achieved in the parabrochi. On the contrary, a reverse flow from the anterior to the posterior side arose if the pressure difference over the parabronchi tended largely towards being negative for a long time.

It was reported that some birds have a narrowing, termed the segmentum accelerans, in the primary bronchus, just cranial to ventrobronchial junction [8,9]. This narrowing is believed to increase convective inertia of inhaled air, thereby directing the inhaled air to the mesobronchi [8,10-12]. In addition, the narrowing causes flow separation at the ventrobronchial junction, which produces vortices near the origin of the ventrobronchi. The vortices formed impede the inhaled air flowing to the ventrobronchi and helps it head towards the mesobronchi [13-15]. These two factors work as aerodynamic valves to direct the inhaled air to the mesobronchi and thereby facilitate the unidirectional flow in the parabronchi throughout the respiratory cycle [8,10-15]. The present model does not possess the narrowing at the primary bronchi. Consequently, inhaled air passing through the primary bronchi was not accelerated to the extent that part of the flow was directed to the ventrobronchi, as seen in Figures 4-6. Moreover, the vortices formed near the origin of the ventrobronchi did not develop significantly, because of absence of the narrowing. If a narrowing was established in the primary bronchus just cranial to the ventrobronchial junction, the inhaled air was more prone to flow to the mesobronchi and subsequently into the posterior air sacs. As a result, it is expected that a pressure difference between the posterior and anterior air sacs becomes larger, thereby achieving unidirectional flow in the parabronchi more easily.

The present study has some limitations. First, the model geometry was an anatomically simplified avian lung and differed from the real lung in the number of bronchi, their angulation, and their diameters [1,2,32,33]. However, the simplification was necessary because of the both limited availability of detailed anatomical information and limited computational resources. In fact, similar geometrical simplifications for flow analyses have been used in many studies, and they successfully captured macroscopic flow characteristics in the avian lung [5,10,12]. In this sense, the usage of the geometrically simplified model would be allowed as far as we discuss only the main features of the flow. Second, mechanical parameters such as flow rates were taken from various species of birds. Thus, the simulation here was a kind of "chimeric" analysis, although we attempted to use data from species with similar body sizes, as far as possible. Third, the model for expressing the mechanics of air sacs to determine the pressure boundary did not take account of intra-thoracic pressure and the elasticity of the air sacs. Here, we imposed an inlet flow condition regardless of the air sac compliance, meaning that breathing was involuntary. Inspiration and expiration should be driven by the intra-thoracic pressure or the transmural pressure over the air sacs. Another problem when modeling the air sacs is the use of a constant compliance value throughout the respiratory cycle. In reality, compliance varies with a change in air sac volume [37,41]. Notwithstanding these limitations, this preliminary study addressed the importance of air sac compliance in avian respiratory function, thereby suggesting that air sac compliance in the avian lung is optimized for each species to achieve unidirectional flow in the parabronchi. In future studies, we will address the shortcomings listed above by including more realistic geometry of the avian lung using magnetic resonance imaging and computed tomography and establishing a realistic respiratory system by incorporating a change in intra-thoracic pressure and air sac elasticity for quantitative analysis of the compliance of avian air sacs.

CONCLUSION

In this study, we examined the effects of air sac compliance on flow in the parabrochi using a geometrically simplified model of an avian respiratory system. The results demonstrate that air sac compliance has a significant impact on flow patterns, including vortex formation, in the lung. Differences in compliance also cause changes in the pressures in the anterior and posterior air sacs and thus the pressure difference between them that drives the flow in the parabronchi. Consequently, it was found that achievement of unidirectional flow in the parabronchi is affected by air sac compliance. These results suggest that air sac compliance would be optimized to accomplish unidirectional flow in the parabronchi.

REFERENCES

1. Akester, A. (1960) The comparative anatomy of the respiratory pathways in the domestic fowl (Gallus domesticus), pigeon (Columba livia) and domestic duck (Anas platyrhyncha). Journal of Anatomy, 4, 487-505.
2. Bouverot, P. and Dejours, P. (1971) Pathway of respired gas in the air sacs-lung apparatus of fowl and ducks. Respiration Physiology, 13, 330-342. doi:10.1016/0034-5687(71)90037-5
3. Brackenbury, J. (1972) Lung-air-sac anatomy and respiratory pressures in the bird. The Journal of Experimental Biology, 57, 543-550.
4. Brackenbury, J. and Amaku, J. (1990) Respiratory responses of domestic fowl to hyperthermia following selective air sac occlusions. Experimental Physiology, 75, 391-400.
5. Brackenbury, J., Darby, C. and El-Sayed, M. (1989) Respiratory function in exercising fowl following occlusion of the thoracic sir sacs. The Journal of Experimental Biology, 145, 227-237.
6. Bretz, W.L. and Schmidt-Nielsen, K. (1972) The movement of gas in the respiratory system of the duck. The Journal of Experimental Biology, 56, 57-65.
7. Brown, R., Kovacs, C., Butler, J., Wang, N., Lehr, J. and Banzett, R. (1995) The avian lung: Is there an aerodynamic expiratory valve? The Journal of Experimental Biology, 198, 2349-2357.
8. Calder, W. and Schmidt-Nielsen, K. (1967) Temperature regulation and evaporation in the pigeon and the roadrunner. American Journal of Physiology, 213, 883-889.
9. Cohn, J. and Shannon, R. (1968) Respiration in unanesthetized geese. Respiration Physiology, 5, 259-268. doi:10.1016/0034-5687(68)90064-9
10. Fedde, M.R. (1998) Relationship of structure and function of the avian respiratory system to disease susceptibility. Poultry Science, 77, 1130-1138.
11. Ficken, M.D. and Barnes, H.J. (1989) Acute airsacculitis in turkeys inoculated with Pasteurella multocida. Veterinary Pathology, 26, 231-237.doi:10.1177/030098588902600307
12. Fletcher, O.J., Anderson, D.P. and Kleven, S.H. (1976) Histology of air sac lesions induced in chickens by contact exposure to Mycoplasma synoviae. Veterinary Pathology, 13, 303-314.

13. Fletcher, O.J., Fairchild, S.E., Smith, F.G. and Trampel, D.P. (1984) Microcomputer-assisted morphometric analysis of airsacculitis caused by Mycoplasma gallisepticum in immunised and nonimmunised chickens. Avian Pathology, 13, 545-560.doi:10.1080/03079458408418554
14. Gilbert, P.W. (1939) The avian lung and air-sac system. The Auk, 56, 57-63.doi:10.2307/4078361
15. Jones, J.H., Effmann, E.L. and Schmidt-Nielsen, K. (1985) Lung volume changes during respiration in ducks. Espiratory Physiology, 59, 15-25.
16. King, A.S. (1966) Structural and functional aspects of the avian lungs and air sacs. International Review of General and Experimental Zoology, 2, 171-267.
17. Kuethe, D. (1988) Fluid mechanical valving of air flow in bird lungs. The Journal of Experimental Biology, 136, 1-12.
18. Maina, J.N. (2005) The lung air-sac system of birds: Development, structure and function. Springer, Heidelberg.
19. Maina, J.N. (2007) Spectacularly robust! Tensegrity principle explains the mechanical strength of the avian lung. Respiratory Physiology & Neurobiology, 155, 1-10.doi:10.1016/j.resp.2006.05.005
20. Maina, J.N. and Africa, M. (2000) Inspiratory aerodynamic valving in the avian lung: Functional morphology of the extrapulmonary primary bronchus. The Journal of Experimental Biology, 203, 2865-2876.
21. Maina, J.N., Singh, P. and Moss, E. (2009) Inspiratory aerodynamic valving occurs in the ostrich, struthio camelus lung: A computational fluid dynamics study under resting unsteady state inhalation. Respiratory Physiology & Neurobiology, 169, 262-270.doi:10.1016/j.resp.2009.09.011
22. Makanya, A. and Djonv, V. (2008) Development and spatial organization of the air conduits in the lung of the domestic fowl, Gallus gallus variant domesticus. Microscopy Research and Technique, 71, 689-702. doi:10.1002/jemt.20608
23. McLelland, J. (1991) A color atlas of avian anatomy. Saunders Co., Philadelphia, 95-119.
24. Onuk, B., Hazitoglu, M. and Kabak, M. (2009) Gross anatomy of the respiratory system in goose (Anser anser domesticus): Bronchi and sacci pneumatici. Ankara Universitesi Veteriner Fakultesi Dergisi, 56, 165-170.

25. OpenFOAM User Guide (2011) Version 2.0.0.
26. Perry, S.F. and Duncker, H.R. (1980) Interrelationship of static mechanical factors and anatomical structure in lung evolution. Journal of Comparative Physiology B: Biochemical, Systemic, and Environmental Physiology, 138, 321- 334. doi:10.1007/BF00691567
27. Russel, S.M. (2003) The effect of airsacculitis on bird weights, uniformity, fecal contamination, processing errors, and populations of Campylobacter spp. and Escherichia coli. Poultry Science, 82, 1326-1331.
28. Sakai, E., Watanabe, T. and Himeno, T. (2005) Oscillatory flow structure and unidirectional flow in model avian bifurcation. Transactions of the Japan Society of Mechanical Engineers Series B, 71, 2083-2091 (in Japanese).
29. Sakai, E., Watanabe, T. and Himeno, T. (2006) Numerical study on mass transport enhancement in oscillatory flow of avian lung model. Transactions of the Japan Society of Mechanical Engineers Series B, 72, 72-80 (in Japanese).
30. Sakai, E., Watanabe, T. and Himeno, T. (2010) Numerical study on oscillatory flow in multi-bifurcation of avian lung model. Transactions of the Japan Society of Mechanical Engineers Series B, 76, 86-93 (in Japanese).
31. Scheid, P. and Piiper, J. (1969) Volume, ventilation and compliance of the respiratory system in the domestic fowl. Respiration Physiology, 6, 298-308. doi:10.1016/0034-5687(69)90029-2
32. Scheid, P. and Piiper, J. (1971) Direct measurement of the pathway of respired gas in duck lungs. Respiration Physiology, 11, 308-314. doi:10.1016/0034-5687(71)90004-1
33. Scheid, P. and Piiper, J. (1989) Respiratory mechanics and air flow in birds. In: King, A.S. and McLelland, J., Eds., Form and Function in Birds, Academic Press, London, 369-391.
34. Schmidt-Nielsen, K. (2007) Animal physiology: Adaptation and environment. 5th Edition, University of Tokyo Press, Tokyo, 37-42 (in Japanese).
35. Shankar, B.P. (2008) Common respiratory diseases of poultry. Veterinary World, 7, 217-219.
36. Tucker, V. (1968) Respiratory physiology of house sparrows in relation to high-altitude flight. The Journal of Experimental Biology, 48, 55-66.

37. Wang, N., Banzett, R., Butler, J. and Fredberg, J. (1988) Bird lung models show that convective inertia effects inspiratory aerodynamic valving. Respiration Physiology, 73, 111-124. doi:10.1016/0034-5687(88)90131-4
38. Wang, N., Banzett, R., Ntions, C. and Jenkins, F. (1992) An aerodynamic valve in the avian primaty bronchus. The Journal of Experimental Zoology, 262, 441-445.doi:10.1002/jez.1402620411
39. West, J., Watson, R. and Fu, Z. (2007) The human lung: Did evolution get it wrong? European Respiratory Journal, 29, 11-17. doi:10.1183/09031936.00133306
40. Womersley, J.R. (1955) Method for the calculation of velocity, rate of flow and viscous drag in arteries when the pressure gradient is known. The Journal of Physiology, 127, 553-563.
41. Yokota, T., Shibahara, T., Wada, Y., Hiraki, R., Ishikawa, Y. and Kadota, K. (2004) Aspergillus fumigatus infection in an ostrich (Struthio camelus). The Journal of Veterinary Medical Science, 66, 201-204. doi:10.1292/jvms.66.201

CITATION

Urushikubo, A. , Nakamura, M. and Hirahara, H. (2013) Effects of air sac compliances on flow in the parabronchi: Computational fluid dynamics using an anatomically simplified model of an avian respiratory system. Journal of Biomedical Science and Engineering, 6, 483-492. doi: 10.4236/jbise.2013.64061.

CHAPTER 4

Hydrodynamic Simulation of Cyclone Separators

R. Utikar[1], N. Darmawan[1], M. Tade[1], Q. Li[1], G. Evans[2], M. Glenny[3] and V. Pareek[1]

[1]Department of Chemical Engineering, Curtin University of Technology, Perth, WA 6845, Australia

[2]Centre for Advanced Particle Processing, University of Newcastle, Callaghan, NSW 2308, Australia

[3]BP Kwinana Refinery Pty Ltd, Mason Road, Kwinana, WA 6167, Australia

INTRODUCTION

Cyclone separatorsare commonly used for separating dispersed solid particles from gas phase. These devices have simple construction; are relatively inexpensive to fabricate and operate with moderate pressure losses. Therefore, they are widely used in many engineering processes such as dryers, reactors, advanced coal utilization such as pressurized and circulating fluidized bed combustion and particularly for removal of catalyst from gases in petroleum refinery such as in fluid catalytic cracker (FCC). Despite its simple operation, the fluid dynamics and flow structures in a cyclone separator are very complex. The driving force for particle separation in a cyclone separator is the strong swirling turbulent flow. The gas and the solid particles enter through a tangential inlet at the upper part of the cyclone. The tangential inlet produces a swirling motion of gas, which pushes the particles to the cyclone wall and then both phases swirl down over the cyclone wall. The solid particles leave the cyclone through a duct at the base of the apex of the inverted cone while the gas swirls upward in the middle of the cone and leaves the cyclone from the vortex finder. The swirling motion provides a centrifugal force to the particles while turbulence disperses the particles in the gas phase which increases the possibility of the particle entrainment. Therefore, the performance of a cyclone separator is determined by the turbulence characteristics and particle-particle interaction.

Experimental and numerical studies have been carried out in the last few decades to develop a better understanding of the flow field inside the cyclone separators. In the early years, empirical models were built (e.g. Shepherd &Lapple, 1939; Lapple, 1951; Barth, 1956; Tengbergen, 1965; Sproul, 1970;Leith &Licht, 1972; Blachman & Lippmann, 1974; Dietz, 1981 and Saltzmann, 1984) to predict the performance of industrial cyclones. However, these models were built based on the data from much smaller sampling cyclones therefore; they could not achieve desired efficiency on industrial scales as the industrial cyclone operates in the turbulent regime while sampling cyclones operate under the transitional conditions. One of the major drawbacks of these empirical models is the fact that they ignore two critical factors that determine the performance of a cyclone namely the unsteadiness and asymmetry. Many flow phenomena such as high turbulence, flow reversal, high vorticity, circulating zones and downflow also occur. The empirical models do not include these phenomena in their analysis and hence are limited in their application. Computational fluid dynamics (CFD) models on the other hand can accurately capture these aspects and thus can take a significant role in analyzing the hydrodynamics of cyclone separators. A validated CFD model can be a valuable tool in developing optimal design for a given set of operating conditions. However, cyclone separators pose a peculiar fluid flow problem. The flow in cyclone separators is characterized by an inherently unsteady, highly anisotropic turbulent field in a confined, strongly swirling flow. A successful simulation requires proper resolution of these flow features. Time dependent turbulence approaches such as large eddy simulation (LES) or direct numerical simulation (DNS) should be used for such flows. However, these techniques are computationally intensive and although possible, are not practical for many industrial applications. Several attempts have been made to overcome this drawback. Turbulence models based on higher-order closure, like the Reynolds Stress Model, RSM, along with unsteady Reynolds averaged Navier - Stokes (RANS) formulation have shown reasonable prediction capabilities (Jakirlic&Hanjalic, 2002). The presence of solids poses additional complexity and multiphase models need to be used to resolve the flow of both the phases.

In this chapter we review the CFD simulations for cyclone separators. Important cyclone characteristics such as the collection efficiency, pressure and velocity fields have been discussed and compared with the experimental data. Several significant parameters such as the effect of geometrical designs, inlet velocity, particle diameter and particle loading, high temperature and pressure have also been analysed. The chapter discusses peculiar features of the cyclone separator and analyses relative performance of various models. Finally an example of how CFD can be used to investigate the erosion in a cyclone separator is presented before outlining general recommendations and future developments in cyclone design.

BASIC DESIGN OF CYCLONE SEPARATORS

A cyclone separator uses inertial and gravitational forces to separate particulate matter from gas. Accordingly various designs have been proposed in literature (Dirgo&Leith, 1986). Figure 1 shows a schematic of widely used inverse flow cyclone and depicts main parts and dimensions. The particle laden gas enters the cyclone separator with a high rotational velocity. Different inlet configurations like tangential, scroll, helicoidal and axial exist to provide high rotational velocity. Of these, the tangential and scroll configurations are most frequent. The rotational flow then descends near the wall through the cyclone body and conical part until a reversal in the axial velocity making the gas flow in the upwards direction. Where this occurs is called as the vortex end position. The upward rotating flow continues along the cyclone axis forming a double vortex structure. The inner vortex finally leads the flow to exit through a central duct, called the vortex finder. The vortex finder protrudes within the cyclone body. It serves both in shielding the inner vortex from the high inlet velocity and stabilizing its swirling motion. The solids are separated due to the centrifugal force and descend helicoidally along the cyclone walls and leave the equipment through the exit duct.

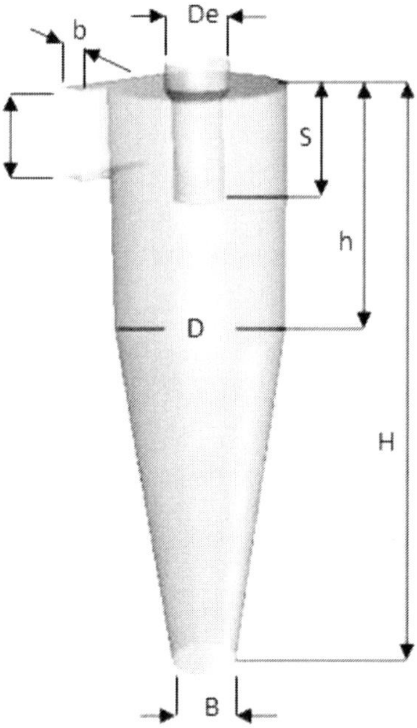

Figure 1: Typical design of cyclone separator.

Table 1: Standard Geometrical Design of Industrial Cyclone Separator

Source	Stairmand (1951)	Stairmand (1951)	Lapple (1951)	Swift (1969)	Swift (1969)	Swift (1969)
Duty	High efficiency	High throughput	General purpose	High efficiency	General purpose	High throughput
D	1	1	1	1	1	1
a/D	0.5	0.75	0.5	0.44	0.5	0.8
b/D	0.2	0.375	0.25	0.21	0.25	0.35
De/D	0.5	0.75	0.5	0.4	0.5	0.75
S/D	0.5	0.875	0.625	0.5	0.6	0.85
h/D	1.5	1.5	1	1.4	1.75	1.7
H/D	4	4	4	3.9	3.75	3.7
B/D	0.375	0.375	0.25	0.4	0.4	0.4

For convenience, the dimensions of various cyclone parts are usually stated in dimensionless form as a ratio to the cyclone diameter, D. This method allows a comparison between the cyclone designs, without using the actual size of each individual part. Table 1 lists a few examples of industrial cyclone types (Leith and Licht, 1972). A more comprehensive range of designs can be found in Cortes and Gil (2007). The performance of a cyclone separator is measured in terms of the collection efficiency defined as the fraction of solids separated and the pressure drop. By nature, the flow in a cyclone separator is multiphase (gas–solid) and shows strong gas–solid–solid interactions. The gas–solid interactions can only be neglected at very low solid loadings. Early CFD models focused on single phase flow and turbulence interactions inside the cyclone. Multiphase CFD simulations that account for the gas–solid and gas–solid–solid interactions and its immediate results concerning cut sizes and grade-efficiency are relatively scarce in the literature. The subsequent sections discuss available CFD models and their predictive capabilities with respect to the flow field, pressure drop and collection efficiency.

COMPUTATIONAL FLUID DYNAMICS MODELS FOR CYCLONE SEPARATORS

The flow inside a cyclone separator is inherently complex and poses many practical difficulties for numerical simulations. The primary difficulty arises from the fact that the turbulence observed in cyclones is highly anisotropic. This renders most of the first order turbulence closures, like the popular k-ε model, unusable for reliable prediction of the flow characteristics. Several attempts were made to overcome this limitation. Boysan et al. (1982, 1983) were one of the first to report CFD studies of cyclone flows. These early studies realized that the standard k-ε turbulence model is not able to accurately simulate this kind of flow and that at least a second-order closure, e.g., RSM is needed to capture the anisotropy and achieve realistic simulations of cyclone flows. The authors found reasonable agreement between the experimental data and simulations using a mixed algebraic-differential, stationary RSM. Many studies have since been performed to capture the turbulence characteristics accurately. The next section will review these in detail.

Selection of numerical parameters, especially the discretization of the advection terms, poses an additional difficulty and plays an important role on the accuracy of simulations. First order discretization is prone to numerical diffusion and often produces misleading results in cyclone separator simulations. The use of hexahedral grids for the main flow region (Harasek et al., 2004) and a second order accurate advection scheme (Bunyawanichakulet al. 2006) has shown a significant improvement in CFD predictions. The flow in a cyclone separator is characterized by unsteady structures like secondary eddies and the precessing vortex core (PVC). An adequate resolution in space and time is necessary to capture these dynamic features. Early CFD studies focused on the steady state solution of the flow (Boysan et al. 1982) due to limited availability of computational power and low spatial resolution that resulted into artificial dampening of instabilities. With increasing computational power, unsteady state simulations with a sufficiently resolved mesh have become standard (Derksen et al. 2006).

Finally, the complexity arises from the presence of solids and their interaction with the gas phase flow. Two approaches, namely the Eulerian-Eulerian approach and the Eulerian-Lagrangian approach have been adopted in the literature to predict the multiphase flow. In the Eulerian–Eulerian approach both the solid particles and the fluid are treated as the interpenetrating continua. The governing equations are then formulated and solved for each phase. This approach can account for the complex phenomena such as the agglomeration and break-up by using a population balance model. The Eulerian-Eulerian approach requires that the interactions between the phases are modelled and are accounted for. These interactions are not yet well understood. The Eulerian-Eulerian approach also requires a specification of the boundary conditions for the particulate phase mutual interaction between particles, and interactions with the wall. In many situations, this information is not readily available. Due to these inherent drawbacks this approach has found limited application in cyclone separator simulation (See for example, Meier et al. 1998 and Qian et al. 2006).

In the Eulerian-Lagrangian approach, particle trajectories are obtained by integrating the equation of motion for individual particles, whereas, the gas flow is modelled using the Navier-Stokes equation. The flow structures in dispersed two-phase flows are a direct result of the interactions between the two phases. Accordingly, a classification based on the importance of the interaction mechanisms has been proposed (Elghobashi, 1994). Depending on the existence of mutual, significant interaction between particles, two different regimes namely dilute and dense two-phase flow can be distinguished (See Figure 2). For $\alpha_p < 10^{-6}$ and $L/d_p > 80$, the influence of particles on the gas can be neglected. This is known as "one-way coupling". The influence of the particle phase is pronounced at higher volume fractions and has to be accounted for. This is known as "two way coupling". For larger particles at higher volume fraction ($\alpha_p > 10^{-3}$, $L/d_p < 8$), the interparticle interactions become important, both through the physical collisions and indirect influence on the nearby flow field. The collisions can lead to coalescence and break-up, which must then be considered. This regime is frequently called the "four-way coupling"regime. The Eulerian-Lagrangian approach is more suited to dilute flows with one- or two-way coupling. The approach is free of numerical diffusion, is less influenced by other errors and is more stable for the flows with large gradients in particle velocity. The treatment of realistic poly-dispersed particle systems is also straightforward. These attributes make Eulerian-Langrangian approach more suitable for the simulation of gas – particle in cyclone separators. The Eulerian-Lagrangian approach is discussed in section 1.3.2.

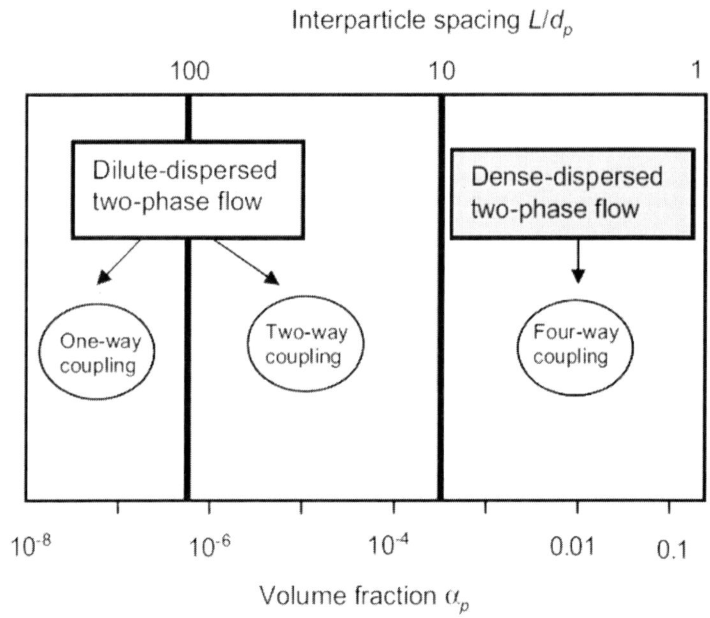

Figure 2: Regimes of dispersed two-phase flow as a function of the particle volume fraction/interparticle spacing. Adapted from Elghobashi, 1994.

Choice of Turbulence Model

The preceding discussion makes it clear that the choice of the turbulence model is the most critical aspect of CFD simulation of cyclone separators. An appropriate turbulence model should be selected to resolve these flow features. As mentioned previously, the models based on first order turbulence closure have a limited ability for capturing the real flow. Generally it is thought that at least a second-order closure is needed to capture the anisotropy and achieve realistic simulations (Hoekstra et al., 1999). While stressing the need for a higher order turbulence model, one needs to keep in mind that as we resolve larger ranges of time and length scales, the computational requirements escalate tremendously. A trade-off between the accuracy and speed of computation is therefore needed for practical simulations.

Of the three available approaches to capture the turbulent characteristics, namely RANS, LES and DNS, RANS approach are the oldest approach to turbulence modeling. In the unsteady RANS, an ensemble averaged version of the governing equations that also includes transient terms is solved. Turbulence closure can be accomplished either by applying the Boussinesq hypothesis, i.e. using an algebraic equation for the Reynolds stresses or by using the Reynolds stress model (RSM), i.e. by solving the transport equations for the Reynolds stresses. In the LES approach, the smaller eddies are filtered and are modeled using a sub-grid scale model, while the larger energy carrying eddies are simulated. The DNS solves fully-resolved Navier - Stokes equations. All of the relevant scales of turbulent motion are captured in direct numerical simulation. This approach is extremely expensive even for simple problems on modern computing machines. Until sufficient computational power is available, the DNS will be feasible only for model problems; leaving the simulation of industrial problems to LES and RANS approaches. Although LES of full-size equipment is possible, it is still costly partly due to the escalating computational cost near the wall region. The unsteady RANS approaches are comparatively far less expensive.

Within the RANS approach, comparative studies have been performed for different turbulence models.Hoekstra et al. (1999) compared the relative performance of the k-ε model, RNG k-ε model (a variation of the k-ε model based on renormalization group theory) and Launder, Reece, Rodi and Gibson (LRRG) models (a differential RSM model). The simulations were compared with Laser Doppler Anemometry (LDA) velocity measurements. Tests were performed with three different vortex finder diameters, which produced three different swirl numbers. The results for the tangential velocity are shown in Figure 3. For all runs, the k-ε model predicted only the inner vertex structure clearly contradicting the experimental observations showed two distinguishing vortices. The RNG k-ε model showed significant improvement, while the RSM exhibited the best behavior. Pant et al. (2002)andSommerfeld and Ho (2003) have also reported similar observations. Gimbun et al. (2005) studied the effect of temperature and inlet velocity on the cyclone pressure drop. They compared four different empirical models, the k-ε model, and the RSM with the experimental data.

Their study of the effect of the inlet velocity on the pressure drop found that the RSM gave the closest agreement with the experimental results. The superiority of the RSM over other models has been established by Meier et al. (1999), Xiang et al. (2005), Qian et al. (2006), Wan et al. (2008) and Kaya et al. (2009). These investigations of various characteristics of cyclone separator flow field, such as velocity profiles, pressure drop, effect of particle size, mass loading, separation efficiency, effect of pressure and temperature, have reemphasized the ability of the RSM for realistic prediction of the flow field inside cyclone separators.

Although, the superiority of the RSM over the other models has been established, it is still not clear which is the most suitable form of the RSM for cyclone separator simulations as both algebraic and differential RSMs have been employed. Between these two, the differential form of the RSM is more accurate and should be preferred over the algebraic form when the extra cost of the calculation is affordable (Hogg &Leschziner, 1989). Within the differential RSMs, the difference between a basic and an advanced differential RSM is also of relevance. For example, Grotjans et al. (1999) compared the predictions of various turbulence models with LDA measurements for the tangential velocity profile in an industrial hydrocyclone. Turbulence models including two differential RSM implementations, the basic Launder, Reece, Rodi (LRR) implementation and the advanced Speziale, Sarkar and Gatski (SSG) implementation along with the standard k-ε and a k-ε model modified to account for the streamline curvature (the k-ε cc model) were tested. They found the flow field to be highly sensitive to the model choice, whereas the pressure distribution predictions were relatively robust. The typical Rankine profile was obtained only by means of the RSMs. The SSG model produced more acceptable results compared to the LRR model in the lower part of the cyclone. The LRR model also underpredicted tangential velocity near the cyclone center.

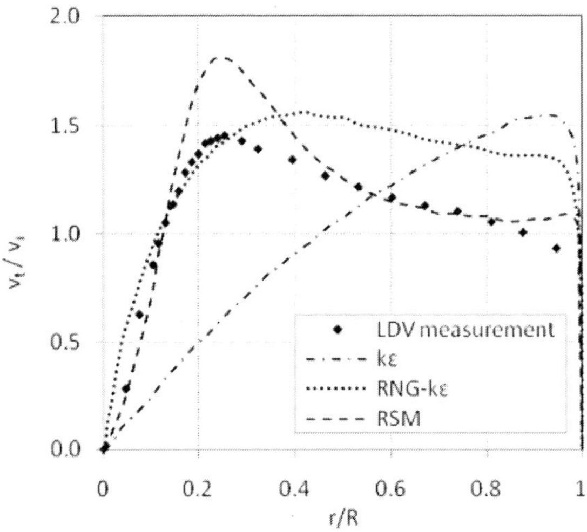

Figure 3: Comparison of tangential velocity profiles (Adapted from Hoekstra et al., 1999).

Despite a number of advances, the ability of unsteady RANS simulations with advanced RSM to accurately predict complex flow structures has not been fully established. Only relatively stable and ordered flows have been simulated. In order to fully establish their viability for cyclone separator simulations, these models should be tested for conditions of highly incoherent and variable PVC. Meanwhile, LES simulation of swirling and cyclone flows is presently becoming a new standard (Derksen, 2008). Derksen and van den Akker (2000) were among the first to simulate the PVC phenomenon by means of LES simulations. The capabilities of LES to simulate the turbulent flow in a cyclone separator have been reported by Shalaby et al. (2005), Derksen (2003), Derksen et al. (2006)and Shalaby et al. (2008). Early simulations (Derksen& van den Akker, 2000) were limited to small scale cyclones at a moderate inlet Reynolds number. With increasing computational power, simulation of industrial scale equipment (with Re = 280000) have been reported (Derksen et al. 2006). The LES approach seems to offer a very realistic simulation. However due to the scale and complexity of today's industrial cyclone separator simulations, the unsteady RANS approach with higher order turbulence closures is the only practical approach that offers affordable realistic predictions of flow inside cyclone. It is only a matter of time that resolved simulations using LES will become the preferred alternative. The behavior of particles and their interaction with continuous phase is paramount in cyclone separators and should be accounted for regardless of the turbulence models.

Eulerian–Lagrangian Approach for Multiphase Flow

The Eulerian-Lagrangian approach is generally more suitable for cyclone separator simulation over the Eulerian-Eulerian approach. In the Eulerian-Lagrangian approach, the gas is treated as a continuum while the solid phase is resolved by tracking individual solid particles (Lagrangian tracking) through the flow field. Lagrangian tracking essentially applied the Newton's second law of motion to a particle to determine its position. The forces generally considered are the drag force, gravity buoyancy, virtual mass and Basset forces. For cyclone separators only the drag and gravity forces are of significance due to the large gas-particle density ratio. Of these, drag force, due to the relative slip between the particle and gas, is the dominating force and is typically modeled using an empirical correlation like the Morsi Alexander model (Morsi& Alexander, 1972).

Depending on the volume fraction, either one-way or two-way coupling is applied to account for the interactions between the two phases. For dilute flows, the gas phase flow is not influenced by the solid flow and Lagrangian tracking decoupled from the gas flow calculation is sufficient. The advantage of this is that the Lagrangian tracking can be performed as a post-processing step as a calculation using the converged and time averaged single phase simulation. To achieve a statically meaningful solution in the simulation, a large number of tracked particles (at least 3×10^5 particles) are required (Sturgess et al., 1985). Furthermore, the time averaged gas flow field smooth out all the turbulent fluctuations. Only particles of large size will behave as exclusively influenced by the time-averaged gas flow. While very small particles will tend to fluctuate following turbulent fluctuations (known as turbulent diffusion) of the gas velocity there will be a complete range of intermediate behaviors between these two extremes. Turbulence fluctuations are random functions of space and time and several ways are available to accommodate them, mostly by including additional terms in the time averaged equation. Amongst these, the stochastic discrete random walk model is the most popular. In this model a prefixed probability distribution of velocity is assumed. The equation of motion for the discrete velocities and particle sizes is solved and the average of the forces is obtained (Gosman&Ioannides, 1981). The Stochastic Lagrangian model has been used successfully by many researchers including Yuu et al. (1978), Boysan et al. (1982), Hoekstra et al. (1999), Sommerfeld and Ho (2003) and Wang et al. (2006). The one-way coupling approach assumes negligible effect of particles on the gas flow. As a result of the collection process, high local solid concentration is observed in the near wall regions. These regions are not effectively modelled using one-way coupling. Hence in most of the Lagrangian tracking results the computational simulations show larger cut-sizes than those observed experimentally.

The effectiveness of the LES to accurately predict the gas flow field in cyclone separators has been established (Derksen, 2003). Subsequently, the Lagrangian tracking has been applied to calculate the particle flow in the LES simulations (Narsimha et al., 2006). The dynamic nature and enormous quantity of time dependent data generated by the LES prohibits post-priori calculation of the Lagrangian tracking and demands instantaneous tracking of a large number of particles. Moreover, since all the turbulence length scales are not fully resolved, the stochastic model for turbulent particle diffusion still needs to be applied. This leads to an extremely intensive method for predicting the cyclone performance. Several alternatives have been proposed, based on average, frozen and periodic LES-velocity fields. Amongst these methods, the periodic approximation produces closest match to experimental data, however, it is the most costly costly method in terms of computational requiremnts. Using this approximation, the simulation results are much closer to the experimental data than the classical Lagrangian tracking (Derksen, 2003).

Depending on the particle size distribution, agglomeration may also become an important factor in predicting the cyclone efficiency. Particle sizes ranging from 1 to 10 μm tend to agglomerate due to the turbulent flow. For this range of particle size the turbulence induced motion is more dominant compared to that of both Brownian motion and gravitational motions. Van der Waal forces are considerably strong enough between the particles to result in particle agglomeration and bigger size particles. Sommerfeld and Ho (2003) observed that the separation efficiency increased considerably for smaller particles in an agglomerating regime. Although, the predictions were not in a perfect agreement with the measurements regarding the grade efficiency curve, they revealed the importance of particle agglomeration on the total separation efficiency.

At higher solid concentrations, the interactions between the two phases become significant and a two-way coupling for the momentum between the particulate and fluid phases needs to be considered. Traditionally, the particle-source-in cell (PSIC) model (Crowe et al. 1977) is used for this purpose. In this model, the flow field is calculated first without the particle-phase source terms until a converged solution is achieved. A large number of ''parcels'' (i.e. discrete particles representing large groups with the same properties) are then tracked through the flow field. The source terms are then obtained from these tracks for a second Eulerian calculation of the gas flow. The procedure is repeated iteratively until convergence is achieved. The accuracy of this method depends on the number of parcels. Typically a minimum of 10000 to 20000 parcels are used. Computational effort also escalates as the number of particles needed to represent the dispersed phase increases. For this reason, two-way coupling therefore is still uncommon. Derksen et al. (2008) studied the effect of mass loading on the gas flow and solid particle motion in a Stairmand high efficiency cyclone separator using a two-way coupled Eulerian-Lagrangian simulation. They observed that compared to one-way coupling the two-way coupled simulation yield higher overall efficiencies. They found that the dependence of the separation efficiency on the inlet solid loading is the result of two competitive two effects namely, the attenuation of the swirl, which lowers the efficiency due to a lowered centrifugal force, and the attenuation of turbulence, which augments the efficiency through a decreased turbulent diffusion of particles.

The standard Lagrangian approach neglects the particle-particle interactions. However at higher solid concentration, these interactions must be included. The discrete element method (DEM) solves the force balance on individual particles and takes into consideration both the particle-particle and particle-gas interactions and has been used to simulate the motion of particles for highly dense flows (Zhu et al. 2007). This approach gives information about the position and velocities of individual particles. Conventional DEM approaches assume a simplified flow field and are not suitable for simulating the particle flow in cyclone separators. Recent advances in DEM and its coupling with the CFD codes has allowed simulation of particle flow within complex flow fields (e.g. Chu et al. 2009), but at this stage the method remains very costly and is limited by the number and size of particles.

Cyclone Flow and Pressure Fields

The collection efficiency and pressure drop performance of the cyclone separator are a direct result of the flow patterns of gas and solid and pressure field inside the cyclone. In a time-averaged basis, the dominant flow feature in a cyclone separator is a vortical flow that can be described as the Rankine vortex, which is a combination of a free outer vortex and a forced inner vortex. Apart from the inlet gas velocity and geometrical parameters, the wall friction and solid loading also influence the strength of the vortex. The empirical models often neglect the later two aspects and hence are limited in their application. Computational modelling is needed to resolve the velocity and pressure fields (Kim et al., 1990, Hoekstra et al., 1999, Ma et al., 2000, Slack et al. 2000 and Solero et al 2002).

Axial Velocity

The axial velocity of the gas phase is a major influence in the transportation of particles to the collection device. Empirical models based on the double vortex structure postulate radially constant values for the downward flow in the outer vortex and upward flow in the inner vortex. Both these values are zero at the axial position where the vortex ends. In reality, the profiles are not flat but exhibit maxima and minima. Typically the downward flow shows a maximum near the walls, while the upward flow shows either a maximum or a minimum at the symmetry axis. The diameter of the swirl of gas entering the vortex finder is larger than the vortex finder diameter itself. Consequently, the gas velocity expected to increase and peak at the vortex finder either on the centre or at the sides. This results in an inverted V or an inverted W shaped profile as seen in figures 4a and 4b for the inner vortex. The V pattern forms an axial velocity maximum at the vortex core of the cyclone while the W pattern forms an axial velocity maximum at the vortex finder radius with a minimum at the vortex core.

Figures 4a and 4b show the axial velocity profile in a cyclone separator at a horizontal position of 0.125 m below the vortex finder for two D_e/D ratios of 0.3 and 0.375, respectively. It can be seen that with relatively small difference in the dimensionless length parameters, the axial velocity in the inner vortex region shift from a V pattern to a W pattern. Harasek et al. (2008) studied this phenomenon by investigating the effect of the vortex finder diameter on the axial velocity profile. Their simulation findings could not determine the conditions when there is a transition in the velocity profile. For smaller D_e/D ratio (< 0.45) the V shaped axial velocity is more stable was dominant. They also observed temporary W patterns due to turbulent fluctuations. At higher D_e/D ratio (> 0.53), the possibility of back flow occurring increases and the air from outside is more likely to be drawn into the core due to the low pressure at the centre of the vortex. Thus, a W pattern is more stable.

The D_e/D ratio is not the only factor that affects the axial velocity profile. The downstream exit conditions, at both gas and particle outlets, have been shown to significantly affect the reversed flow to the vortex finder and the whole internal flow. For example, Hoffmann et al. (1996) showed that for cyclone separators with diplegs, the upward flow has a V-shape profile. This was confirmed by the numerical study by Velilla (2005). The V-shape profile is expected to have greater separation efficiency due to a narrower ascending flow region with higher swirl. Some observation by Wang et al. (2006)and Liu et al. (2006) indicate that the centre of the upward flow does not always coincide with the centre of the geometrical cyclone. This is attributed to the chaotic flow within the cyclone. In such cases, an eccentric vortex finder can be designed to reduce the pressure drop and weaken the chaotic flow.

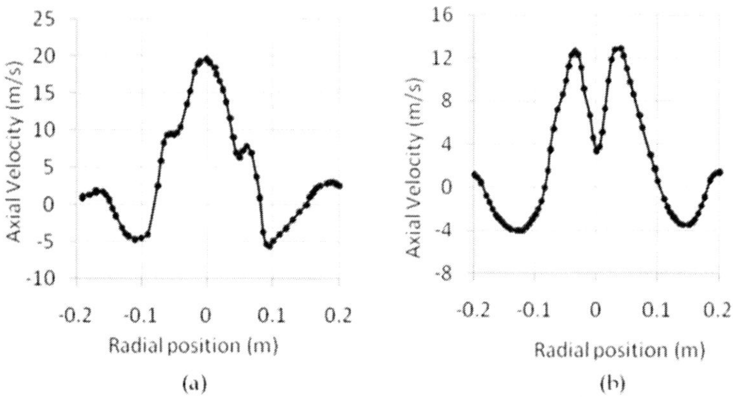

Figure 4: Typical axial velocity profile (a) V pattern and (b) W pattern (Adapted from Harasek et al. 2008).

Radial Velocity

The radial velocity affects the particle bypass and is an important factor in analyzing the particle collection and losses of efficiency. Frequently the radial velocity is assumed to be of much lesser magnitude than the other components. However, this is valid only in the outer vortex, and especially near the vortex finder, the radial velocity increases rapidly towards the vortex core (Muschelknautz, 1972).

A typical example of radial velocity field is shown in Figure 5. The plots are at the center of the cyclone (Figure 5a) and at cut of sections B-B and C-C (Figure 5b). The radial velocity profile at section D-D of the cyclone (Figure 5c) has a helical shape. The axis of the vortex is slightly curved and not aligned with the geometrical axis of the cyclone. It can be seen from Figure 5a that the distribution of the radial velocity is positive on one side and negative at the other side. This is due to the non-symmetrical shape of a conventional tangential inlet cyclone. It is also observed that the radial velocity increases sharply towards the vortex core. Alekseenko (1999) suggests that this phenomenon is the result of vortex rotation along the flow, in a helical shape, around the geometric axis of the cyclone. Points A and D in Figure 5 show a short circuiting flow that can hinder the cyclone separator performance. Point A is called 'lip leakage". It is located below the rim of the vortex finder where a radial component of velocity flows inward directly to the vortex finder instead of flowing down the outside wall and returning in the central core. Point D is located near the inlet duct and the vortex finder. Just outside the vortex finder the radial velocity indicates an inward flow from the inlet (a negative value) but due to the effect of the centrifugal force around the vortex finder its value rapidly changes to zero and becomes positive. This results in an instability in the cyclone and it may affect the cyclone performance.

Tangential Velocity

The flow within a cyclone is dominated by the tangential velocity and strong shear in the radial direction which results in a centrifugal force that determines the particle separation. Subsequently, much discussion within cyclone separator studies is focused on the tangential velocity (Cortes & Gil, 2007).

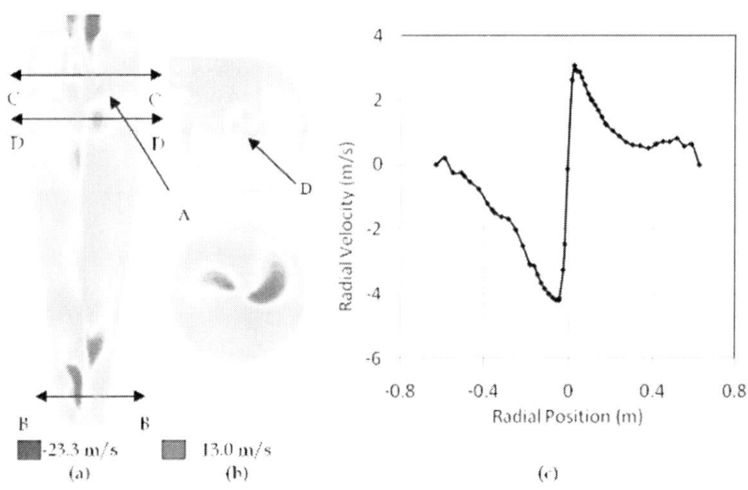

Figure 5: a) Contours of radial velocity at a vertical plane (b) Contours of radial velocity at horizontal cut off section B-B and C-C (c) radial velocity profile at section D-D.

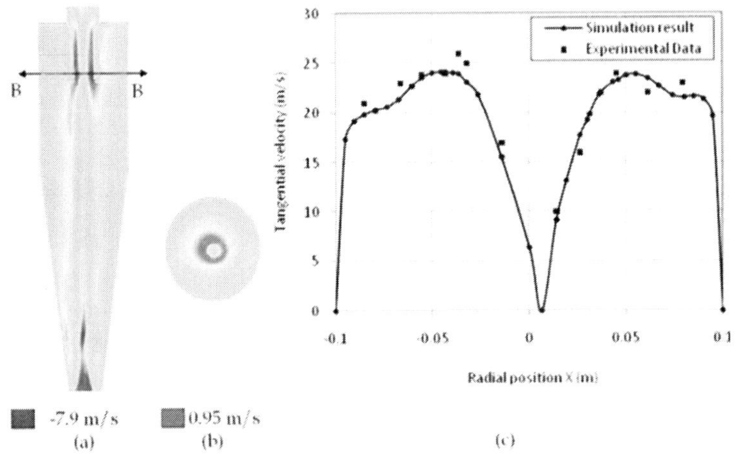

Figure 6: a) Contours of tangential velocity at a vertical plane (b) Horizontal cut off section at Line B-B (c) Example of comparison of numerical and experimental results for tangential velocities (Adapted fromWang et al., 2006).

Typical contour plots of the tangential velocity in both vertical and horizontal planes are shown in figures 6a and 6b, respectively. Figure 6c shows comparison of numerical and experimental results for tangential velocities from Wang et al. (2006). The cyclone has an asymmetrical shape and as can be seen from figure 6a, the axis of the cyclone does not exactly coincide with the axis of the vortex. The Rankine vortex can also be visualized. Figure 6b shows the plot of the tangential velocity across horizontal lines. It is observed that the inlet speed is accelerated up but then it decreases when the gas spins down along the cyclone wall. At a certain point flow reversal takes place and the gas flows in the reverse direction to the exit. Before entering the vortex finder, the gas collides with the follow-up flow and velocity decreases sharply. This causes energy loss and pressure drop. The tangential velocity is highly dependent on the geometrical design, wall friction and particle loading. Wang et al. (2006), Wan et al. (2008) and Raoufi et al. (2009) have demonstrated the use of CFD in reasonably predicting the tangential velocity under varied conditions.

The temperature also has an effect on the tangential velocity (Shi et al., 2006). A minor decrease is noticed at the area of the inner vortex with increasing the temperature. The overall and maximum tangential velocity is also decreases on increasing the temperature. As the gas moving toward the vortex finder, the area of inner vortex become narrower and the outer vortex become wider. The main reasons for the changes are that on increasing the temperature the gas density decreases and viscosity increases. Furthermore, the centrifugal force is proportional to the square of the tangential velocity, therefore higher temperature causes the centrifugal force to decrease hence the lower separation efficiency.

Pressure Field

The pressure drop across the cyclone is a significant variable since it is directly related to the operating costs. The pressure drop is defined as the difference between the static pressure at the inlet and outlets. Conventional tangential inlet cyclone operations induce a spinning motion that creates radial pressure gradients, which provide a curvature for the gas flow. The particles usually follows these trajectories directed toward the cyclone wall. The Pressure drop within a cyclone is contributed to by both local losses and frictional losses. Local losses include the expansion loss at the inlet and the contraction loss at the outlet while the frictional losses include the swirling loss due to gas to wall friction and the dissipation loss of the dynamic energy of gas.

The total pressure drop, comprising both static and dynamic pressure, decreases on increasing the wall friction coefficient, particle concentration and cyclone length. The combination of the static pressure and the kinetic energy of the vortex is called the total pressure. The viscous dissipation of the kinetic energy in the vortex finder dominates the pressure loss within the cyclone (Dirgo, 1988 and Coker, 1993). Thus the pressure loss is directly proportional to the dynamic pressure. About 40% of the pressure drop is due to the swirl energy losses while the rest is from the sudden expansion at the inlet and the contraction at the outlet duct. Any influence that increases the vortex strength will increase the pressure loss. For example, an increase in the wall friction coefficient will increase the pressure loss as it will decrease the velocity magnitude and will lead to decreased loss in the vortex finder. Similarly, an increase in the wall friction will decrease pressure drop (Hoffman et al. 1992). At increased particle concentration, the tangential velocities will be lower and accordingly will yield a lower pressure drop.

Gimbun et al. (2005) studied the effect of the inlet velocity and particle loading on pressure drop. They compared experimental values by Bohnet (1995) with empirical models by Shepperd and Lapple (1939), Casal and Martinez (1983), Dirgo (1988), Coker (1993), as well as with CFD predictions using the k-ε model and the RSM model. The result showed that the RSM model produced the closest pressure drop prediction. The k-ε results showed a reasonably good prediction at about 14% -18% deviation. The CFD studies on gas-solid flow in a cyclone separator by Wang et al. (2006) using the RSM model also showed an acceptable agreement with experimental data for Stairmand high efficiency cycloneHoekstra et al. (1999).

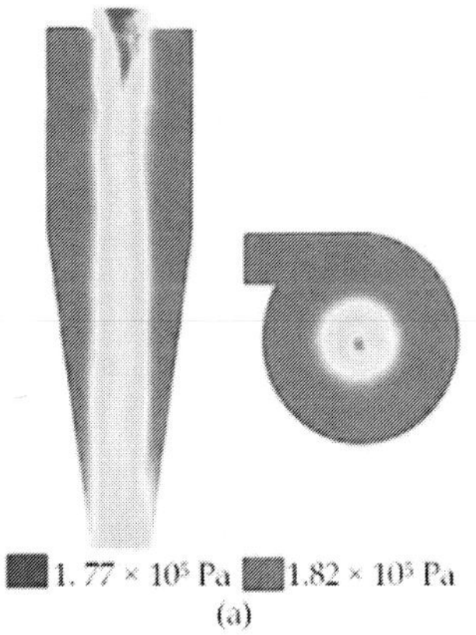

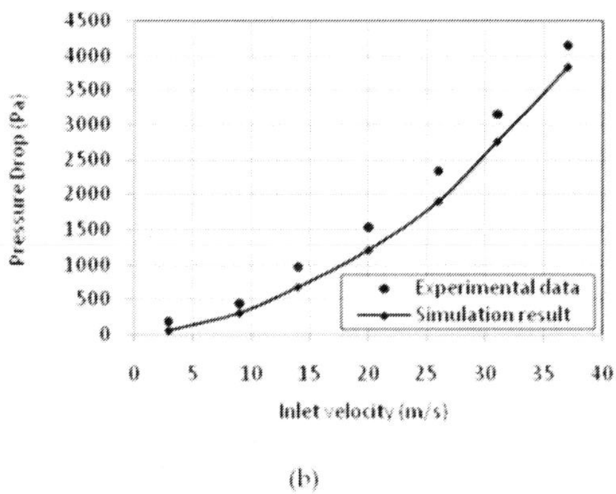

Figure 7: a) Contours of total pressure (b) Pressure drop comparison of CFD RSM measurement and experimental data (Stairmand cyclone) for various inlet gas velocities (Adapted from Wang et al., 2009).

Figures 7a and 7b show the contours of the total pressure and the pressure drop comparison between RSM model predictions and experimental data for the Stairmand cyclone at various inlet gas velocities, respectively. The total pressure increases in the radial direction from the thecentre to the wall of the cyclone. Flow reversal in a cyclone is due to the low pressure centre. It can be seen from Figure 7bthat the CFD simulations underpredict the pressure drop, across the cyclone only the static pressure is considered with the dynamic pressure being neglected. In reality swirl dissipation continues further down the cyclone outlet so that dynamic pressure will be lost without any chance to be recovered. Consequently, the actual pressure drop will be higher.

Any factors that may cause change in the absolute magnitude of the velocity, which in turn changes the strength of the vortex, will affect the pressure drop in the cyclone. Generally, the pressure drop will increase with increasing vortex strength. The pressure drop will decrease with an increase in the wall friction coefficient, particle loading or cyclone length (Yu et al., 1978, Parida& Chand, 1980,Hoffmanet al., 1992). When the wall friction coefficient is increased, the swirl in the separation space decreases and causes an increase in the pressure loss. However, it also decreases the absolute velocity magnitude which results in a decreased pressure loss at the vortex finder. The latter effect is always much higher than the first effect and any increase in the wall friction decreases the pressure drop.

Some important processing industries such as pressurized fluidized bed combustion (PFBC), Integrated gasification and combined cycle (IGCC) and Fluid Catalytic Cracking (FCC) operate at high temperatures and pressures. The operating temperature and pressure will influence the gas density and viscosity and their effect on the drag force. Therefore, for these industries, the operating temperatures and pressures are the important parameters that determine the pressure drop in the cyclone. Shin et al. (2005) (See Figure 8) conducted numerical and experimental study on the effect of temperature and pressure on a high efficiency cyclone separator. They found that the pressure drop decreases at a higher pressure and lower temperature. Higher pressures and lower temperatures increase the gas density which in turn creates a higher dynamic pressure hence the higher pressure drop. This trend is confirmed by similar experimental and numerical studies by Gimbun et al. (2005) and Shi et al. (2006).

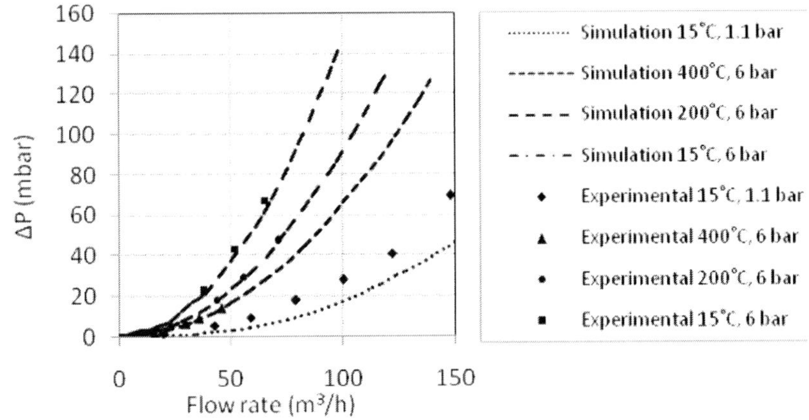

Figure 8: Comparison of experimental and numerical result for pressure drop at a given flow rate in a elevated pressure and temperature (Adapted from Shin et al., 2005).

Collection Efficiency

The fraction of solids separated at the outlet is defined as the collection efficiency. Since cyclone separators usually handle various sizes of particles, the efficiencies are defined according to a continuous narrow interval of particular group size particles. The swirling motion within the cyclone separator causes large particles to travel swiftly to the cyclone walls and roll down to the outlet. On the other hand, the smaller particles are often drifted in upward spiral flow due to the slower speed and escape through the gas outlet. This typically yields an S shaped curve for the collection efficiency. Particle collection is the net effect of various forces acting on the particles. It is well known that the collection efficiency is governed by the centrifugal, gravitational and drag forces (Blachmann& Lipmann, 1974). Factors such as the particle-particle and particle-wall interaction also influence the cyclone efficiency. Their effect is not yet fully understood and hence often neglected in empirical modelling. Further, the empirical models are based on the lab scale data. Depending on the operating conditions, the flow inside the cyclone can be laminar, transitional or turbulent for the lab scale equipment. The actual industrial cyclonesoperate in the turbulent regime where the friction and its corresponding outcomes are significant. Therefore the particle collection efficiency models based on the lab scale data may not accurately predict the collection efficiency for industrial cyclones. At lower mass loading (<5-10 g/m³) the empirical models perform reasonably well (Cortes and Gil, 2007). Many cyclone separator systems of industrial interest such as the FCC, PFBC and CFBC are well known for handling high particle loadings, where, the interphase and interparticle processes become important and the predictive ability of the conventional models is poor. Numerical studies then become necessary to achieve a better understanding of the cyclone collection efficiency.

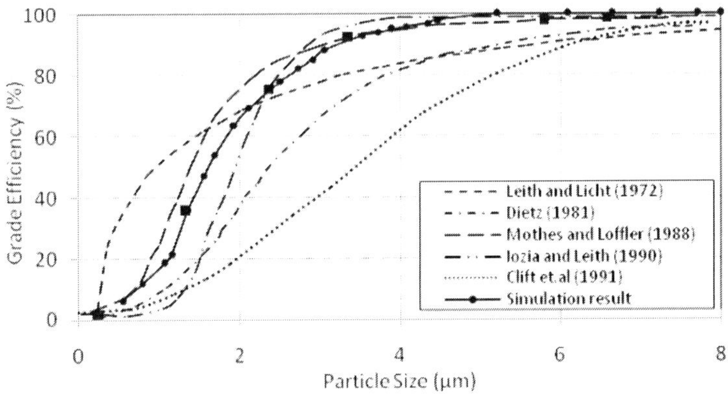

Figure 9: Collection efficiency model comparison: theoretical, numerical and experimental (Adapted from Zhao et al., 2006).

A typical comparison of relative performance of computational models and conventional models in predicting the cyclone efficiency is shown in Figure 9 (Zhao et al. 2006). The collection efficiency in a cyclone with conventional inlet (CI) and a spiral double inlet (DI) configuration is evaluated using the unsteady RANS model with the RSM turbulence model. These predictions are compared with the experimental data and empirical models. Both these profiles follow the S shape with lower collection efficiencies for the smaller particles and almost total capture of larger particles. The comparison also clearly demonstrates the superiority of CFD model over the empirical model in calculating the collection efficiencies. The collection efficiency is a primary measure of the performance of a cyclone separator and depends on the operating conditions and the geometrical characteristics. In the following subsections we look at how these affect the collection efficiencies.

Effect of Mass Loading, Particle Diameter and Inlet Velocity on Cyclone Efficiency

Qian et al. (2006) investigated the effect of mass loading on the collection efficiency. The results of their simulation are shown in Figure 10. The collection efficiency is defined as the ratio of mass flow rate at the inlet and outlet for a converged steady condition. It is clearly evident that the collection efficiency increase on increasing the particle loading. The result is consistent with most of the previous studies (like Stern et al. 1955). Different mass loading for various particle group-sizes affect the grade efficiency differently. Smaller particle group sizes show a higher efficiency increase compared to the larger particle size groups. These findings are also confirmed by the simulation and experimental study by Luo et al. (1999) and Ji et al. (2009). The increase in cyclone efficiency with solid loading is more pronounced at lower gas velocities (Hoffmann et al. 1991, 1992).

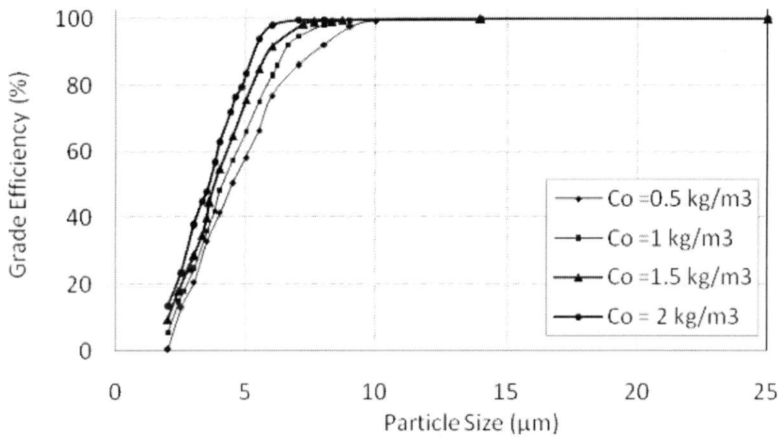

Figure 10: Separation efficiency simulation result for various inlet particle concentration with constant inlet velocity (Adapted from Qian et al., 2006).

Mass loading effect is usually coupled with the particle diameter. At lower mass loadings, the smaller particles (< 10μm) tend to be dispersed and hauled by the gas flow and escape from the vortex finder at the top of the cyclone separator (Derksen 2003 and Wan et al. 2008). But on increasing the particle mass loading, a sweeping effect of the coarser particle that sweeps away the smaller particles to the cyclone wall is observed. The swept particles then roll down and are collected at the bottom of the cyclone. This effect is also responsible for the formation of agglomerates. Agglomeration causes increased centrifugal force on the smaller particles improving their collection efficiency. Wan et al. (2008) also note that on increasing the particle loading, both the downward flow and the axial velocity at the centre (in upward direction) increase. This aids in higher collection efficiency in the cyclone separator.

The inlet gas velocity also has an effect on the collection efficiency. The effect is also tightly related to the particle mass loading. Figure 11 shows the effect of increasing the inlet gas velocity on the collection efficiency for a given solid loading (Ji et al. 2009). The collection efficiency increases with the inlet gas velocity. For smaller particle sizes (< 10 µm), the increase in the efficiency with respect to the gas velocity is more pronounced. As the particle size increases, the effect of the inlet velocity becomes insignificant. These observations are in line with the experimental work of Fassani et al. (2000) and Hoffmann et al. (1991). The higher inlet velocity, results in the higher tangential velocity, thus leading to a higher centrifugal force and collection efficiency. Patterson and Munz (1989)analyzed the effect of several parameters including the gas temperature (300K - 2000K), inlet gas velocities (3 m/s - 42 m/s) and particle loadings (up to 235.2 g/m^3) on the cyclone efficiency. Their analysis showed that there is an increase in the cyclone efficiency especially under high temperature condition.

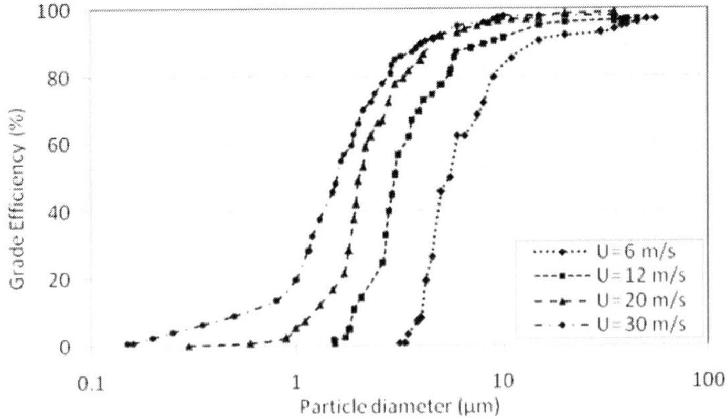

Figure 11: Grade efficiency for different inlet velocity (Adapted from Ji et al., 2009).

Effect of Geometrical Configuration on Cyclone Efficiency

The geometrical configuration is probably the most crucial aspect affecting the performance of a cyclone separator. The cyclone separator performance is sensitive to the smallest of geometrical changes. Through meticulous experimentation, optimal designs of cyclone separators have been proposed. One of the most commonly used designs is the Stairmand high efficiency cyclone (Stairmand 1951). After almost 60 years, this design is still popular since it manages to maintain the velocity profile at every axial point well and achieves very high collection efficiency. The key for Stairmand high efficiency cyclone design lies on the optimal geometrical ratio of cone length to diameter (L/D=4). Unfortunately, it is impossible to always accommodate Stairmand cyclone in the industrial processes, therefore an optimal design best suited for the need has to be explored. Cyclone peripherals like the dipleg and the vortex finder also affect the performance.

Several studies on the effect of dipleg on the cyclone performance have been reported (Hoffmann et al., 1996, Gil et al., 2002, Obermair et al., 2003 and Kaya et al., 2009). Hoffman et al. (1996) found that the L/D ratio of 5.65 is optimal, after this, the collection efficiency starts decreasing. Kaya et al. (2009) extended the effect of dipleg to investigate various inlet velocities and particle diameters for smaller cyclones. They showed that cyclone performance can be improved by adding a correct length of dipleg at the particle outlet. A prolonged cyclone improves the collection efficiency by providing additional space for the separation. According to their results, the optimum dipleg length would be approximately half the height of the cyclone. Placing the vortex end – the location at which the outer vortex changes its direction, exactly at the end of dipleg leads to high collection efficiency. Increasing the length further causes the vortex to fall short and leads to the re-entrainment of particle to the vortex finder thus decreasing the efficiency. Similar observations were made by Gil et al. (2002).

The vortex finder provides another avenue for optimizing the cyclone design. The size of the vortex finder is critical since it determines the inner and outer swirling flow pattern within a cyclone separator. Concrete methodology is not yet available to optimize the vortex finder. Saltzmann et al. (1984) and Iozia et al.

(1989) studied the effect of vortex diameter on the cyclone performance. Kim and Lee (1990) provide information on how the vortex finder and cyclone body diameter affects the cyclone performance. Lim et al. (2004) evaluated different geometries of the vortex finders to optimize the performance. Their results concluded that with smaller vortex finder diameters, the tangential velocity in the inner region of the cyclone increases resulting in a higher collection efficiency. These findings were supported by Raoufi et al. (2009) using CFD modelling.

CFD modelling has opened an avenue for the cost effective optimization of the cyclone geometry. Many geometrical designs have been proposed using CFD studies that can be used to improve the performance of cyclone separators. The list of these modifications include, (i) the use of different inlet types (scroll, helicoidal, axial spiral double inlet and square cyclone inlets) can be found in Cortes and Gil (2007), Wang et al. (1999) and Zhao et al. (2006), (ii) including a long coned (Lee et al., 2006), and (iii) the variation in body and cone height (Xiang et al., 2005).

EROSION IN CYCLONE SEPARATOR

The erosion often poses a serious problem in industrial cyclone separators. It is caused by the friction between the particles and the cyclone walls due to the continuous motion of the particles along the cyclone wall. The erosion often leads to a physical damage to the equipment (holes) and costly unscheduled shutdowns. Even pinholes in the wall can disrupt the pressure balance and can cause increased emissions. The eroded wall material may also contaminate the particles collected at the outlet. The erosion occurs due to directional or random impingement of the solid particles and due to friction of the sliding particles against the cyclone wall. Assuming a perfect elasticity, a particle colliding with the cyclone wall will rebound at the same angle (in opposite direction) as it hits the wall, the erosion is minimal at larger angles, however, at very acute angles of impingement, severe erosion results along the cyclone wall circumference. On a flat and ductile target the greatest erosion occurs at around 15 impingement angle (Noppenberger 2000). The erosion is proportional to the kinetic energy of the particles hitting the cyclone wall and the intrinsic strength of the particles. It also depends upon the angle of incidence on the cyclone wall.

At certain specific locations significantly higher erosion occurs compared to other locations, most possibly due to high local velocity which is a function of the particle size and particle density. The high erosion rates are also observed at the inlet. Severe erosion is also observed inside the cylone, however, the there is no consensus on the location of this. Jones et al. (1979) found that the area extending 180° from the inlet point has most severe erosion and the erosion maxima is observed at angular positions of 105°, 165°, 205° and 245° from the tangential inlet. Another study by Youngdahl (1984) found that the highest rate of erosion occurs at 28 and near the lower edge of the inlet stream. Because a detailed velocity and pressure fields and particle trajectories are available from an Eulerian-Lagrangian CFD simulation, it can be used to determine the otherwise impossible locations for likely erosion sites.

Figure 12 shows the erosion map for an industrial cyclone using (a) uniform particle size of 80μm at the inlet and (b) Rosin – Ramler particle size distribution at the inlet with average particle size of 80μm. The results show that the particles follow a distinct path rather than cluttering on the cyclone walls while swirling down. For the uniform particle size, most severely eroded sites were the top part of the cyclone cylinder (near the inlet) and the cone of the cyclone. This observation is in accordance with the experimental findings of Jones et al. (1979) and Yongdahl et al. (1984). On using a particle size distribution, the most eroded part in the cyclone was at the intersection between the cyclone cylinder and cone. This demonstrates the sensitivity of the erosion to particle size distribution.

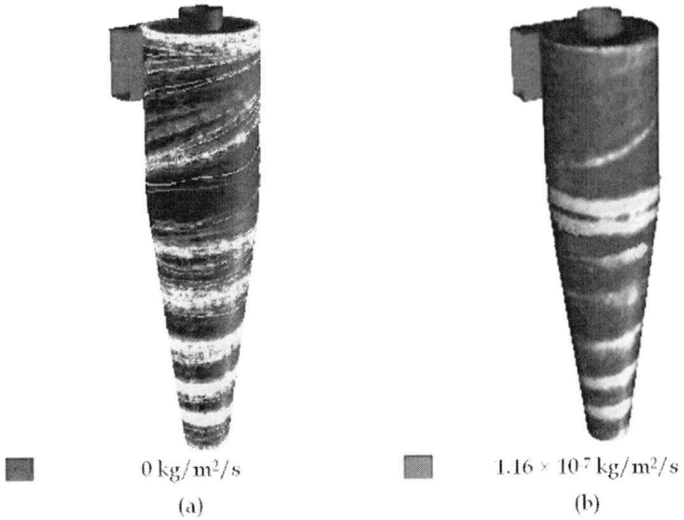

Figure 12: a) Contours of erosion (a) with uniform catalyst size (b) Rosin-Rammler PSD.

Effect of Particle Size Distribution on Cyclone Wall Erosion

To investigate the effect of particle size distribution, a wider inlet particle size distribution (1 µm to 160 µm) was simulated using the Lagrangian tracking. The resulting particle trajectories are shown inFigure 13. It is clearly evident that the larger particles are collected at the walls while the smaller particles escape downwards in spiral manner. This is because the drag force on the smaller particles is larger than the centrifugal force preventing their transportation to the cyclone wall. The results show that the particle with sizes < 40 µm escape from the bottom of the cyclone after a certain residence time while the particle with sizes > 60 µm keep spinning around the mid level for significantly longer.Wang et al. (2006) have shown this phenomenon experimentally using ceramic balls. One possible explanation lies in the balance of the centrifugal force versus the gravitational force. As the larger particles roll down the conical part, the centrifugal force on the particle increases because the radius of the cyclone decreases while the tangential velocity of the particle remain almost the same. The larger particles will eventually be collected at the solid outlet due to the particle-particle interaction. However, some of the particles will stay at the cyclone wall. There also appears to be a critical value of the particle diameter below which the particle is not expected to be collected at the outlet. The critical value of particle diameter is related to the cyclone geometry, gas inlet velocity and particle properties.

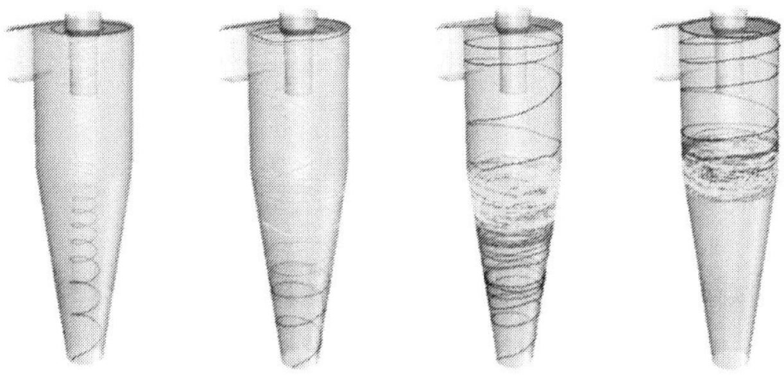

Figure 13: Single particle trajectories of size (a) 5 μm b) 20 μm (c) 40 μm (d) >60 μm.

Effect of Particle Mass Loading on Cyclone Wall Erosion

The effect of particle mass loading on the average erosion rate at different gas velocities is shown inFigure 14. The average of erosion rate gives a global measure of this quantity within the entire cyclone. As expected, at a given gas velocity, the erosion rate decreases with the solid loading. For a given solid loading, however, the average erosion rate increases with gas velocity upto a certain velocity. Beyond this velocity, the erosion rate remains constant or decreases slightly.

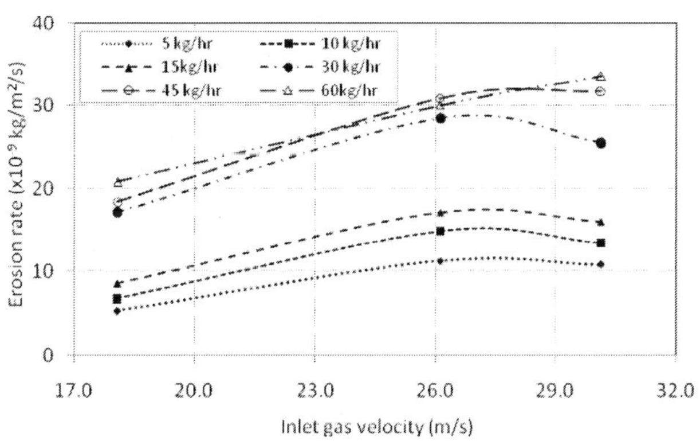

Figure 14: Comparison of erosion rate with various gas and particle flow rate.

At lower gas velocities, lower momentum is imparted by the gas on the particle, which sometimes prevents a second rebound to happen and the particle is forced to stay near the cyclone wall. Consequently, the rate of erosion is lesser at lower velocities. As the gas velocity is increased, the particle rebound is more likely to happen. Since the particle rebounds are not perfectly elastic, it reduces the impact angle gradually after the first rebound, thus increasing the average erosion rate. At even higher gas velocities, the centrifugal force on the particles increases. This makes some particles reach the cyclone wall faster. As a consequence, a layer of slow moving particles is formed that protects the walls from collision by other particles which then reduces the rate of erosion slightly.

SUMMARY

Due to their simple and robust construction, cyclone separators are widely used in the chemical and process industries. In spite of their simple construction, flow patterns inside cyclones are highly complex and not fully understood. Understanding the flow is critical in accessing their performance and CFD is a useful tool in providing this information. However, due to the very nature of the flow, the application of CFD should be exercised with prudence. In order to accurately resolve the unsteady nature of the flow inside a cyclone, higher order numerical discretization along with unsteady simulation (unsteady RANS or LES) are needed. It also requires a higher order turbulence model (atleast second order like RSM) for the unsteady RANS simulations. For resolving the gas flow field, the LES provides superior results than the RANS approach. However, the cost of LES is prohibitive for the industrial-scale devices. Recent developments turbulence modelling such as the differential RSMs have shown a light of hope to achieve LES level of accuracies at RANS cost. However, the conditions under which the unsteady RANS solver can be used in place of the LES need to be explored.

To obtain the particle flow, the Eulerian-Lagrangian formulation with either one way or two way coupling should be employed. The necessity of capturing the unsteadiness of the gas flow in combination to the flow of poly-disperse particulates demands far superior computational power than what is currently available. Therefore, a considerable progress needs to be made in multiphase flow simulation of cyclones. Factors like interparticle phenomena and the boundary condition at the wall also need a careful attention. The advances in the DEM-CFD coupled simulations will bring new insight in the calculation of highly-loaded cyclones. Nevertheless, the two phase flow simulations have provided some useful insight into the cyclone operations and have provoked to question the existing theories on the particle flow and separation in cyclones.

The enormous computational requirements, even for the minimal modelling of cyclone phenomena, have limited our ability to go beyond a simple understanding of the flow structures, collection efficiency and global design issues. More systematic research for addressing the other important issues, such as reasonable estimates of the cyclone natural length and vortex finder dimensions, is needed. Furthermore, the loss of coherence in the vortex and the ensuing chaotic flow patterns and the effect of swirl-stabilization are some of the other topics that remain unanswered. With increasing computational power, it is envisaged that in the near future we will be able to perform fully resolved simulations on cyclones which will not only answer the above questions but will also advance our knowledge of cyclone operations and even optimize them for specific operational circumstances.

REFERENCES

1. W. Barth, 1956 Design and Layout of the cyclone separator on the basis of new investigation. Brenn. Warme Kraft, 8 1 9 .
2. M. W. Blachmann, M. Lippmann, 1974 Performance characteristic of the multi cyclone aerosol sampler, IndHyg. Ass. J., 35 311 326 .
3. M. Bohnet, 1995 Influence of the Gas Temperature on the Separation Efficiency of Aerocyclones, Chem. Eng. Process, (34), 151.
4. F. Boysan, W. H. Ayers, J. Swithenbank, 1982 A fundamental mathematical modeling approach to cyclone design, Trans. Isntn. Chem. Engrs, 60 222 230 .

5. F. Boysan, B. C. R. Ewan, J. Swithenbank, W. H. Ayers, 1983 Experimental and theoretical studies of cyclone separator aerodynamics, InstChemEngSymp Ser,69 305 319 .
6. P. Bunyawanichakul, M. P. Kirkpatrick, J. E. Sargison, G. J. Walker, 2006 Numerical and Experimental Studies of the Flow Field in a Cyclone Dryer, J Fluids Eng, 128 (6), 1240
7. J. M. Casal, A. Martinez, 1983 A better way to calculate cyclone pressure drop. Chemical engineering., 90, 99.
8. K. W. Chu, B. Wang, A. B. Yu, A. Vince, 2009 CFD-DEM modeling of multiphase flow in dense medium cyclone. Powder Tech., 193(3), 235-247.
9. K. Coker, 1993 Understand cyclone design. Chemical Engineering Progress, 28 51 55 .
10. Cortes, A. Gil, 2007 Modeling the particle flow inside cyclone separators. Progress in energy and combustion Science (33).
11. J. J. Derksen, H. E. A. van den Akker, 2000 Simulation of vortex core precession in a reverse flow cyclone. AIChE J., 49 1359 1371
12. J. J. Derksen, 2003 Separation performance predictions of a Stairmand high-efficiency cyclone. AIChE J (49), 1359 1371 .
13. J. J. Derksen, 2005 Simulations of confined turbulent vortex flow. Comp Fluids (34), 301 318.
14. J. J. Derksen, S. Sundaresan, H. E. A. van der Akker, 2006 Simulation of mass loading effects in gas-solid cyclone separators. Powder Tech., 163 59 68 .
15. P. W. Dietz, 1981 Collection efficiency of cyclone separators. A.I. Ch.E.J, 27, 888 892 .
16. J. Dirgo, 1988 Relationships between cyclone dimensions and performance. Havarad University, USA.
17. J. Dirgo, D. Leith, (Ed.) 1986 Encyclopedia of fluid mechanics, (Vols. 4). Houston: Gulf Publishing Company.
18. S. E. Elghobashi, 1994 On predicting particle-laden turbulent flows. ApplSci Res; (52), 309-29.
19. Gil, C. Corte´s, L. M. Romeo, J. Velilla, 2002 Gas-particle flow inside cyclonediplegs with pneumatic extraction. Powder Tech. (128), 822-829.
20. J. Gimbun, T. G. Chuah, A. Fakhru'l-Razi, T. S. Y. Choong, 2005 The influence of temperature and inlet velocity on cyclone pressure drop: A CFD study. Chemical engineering and Processing, 44 7 12 .

21. D. Gosman, E. Ioannides, 1981 Aspect of computer simulation of liquid fuelled combusters. AIAA paper, 81-0323.
22. H. Grotjans, F. R. Menter, R. C. Burr, M. Gluck, 1999 "Higher Order Turbulence Modelling in Industrial Applications", 4th Int. Symp. on Engineering Turbulence Modeling and Measurement..
23. M. Harasek, A. Horvath, C. Jordan, 2004 Investigation of dependence of gas flow on the geometry of cyclonic separators by CFD simulation. Paper presented at the CHISA 2004 - 16th International Congress of Chemical and Process Engineering.
24. M. Harasek, A. Horvath, C. Jordan, 2008 Influence of vortex finder diameter on axial gas flow in simple cyclone. Chemical product and process modeling 3(1), Article 5.
25. J. Hoekstra, J. J. Derksen, H. E. A. Van Den Akker, 1999 An experimental and numerical study of turbulent swirling flow in gas cyclones. Chem. Eng. Sci., 54 2055 2056 .
26. C. H. Hoffmann, H. Arends, H. Sie, 1991 An experimental investigation elucidating the nature of the effect of solids loading on cyclone performance, Filtration & Separation, 28 (3), 188 193
27. C. Hoffmann, A. van Santen, R. W. K. Allen, R. Cliff, 1992 Effects of geometry and solid loadings on the performance of gas cyclones. Powder Tech., 70 83 91 .
28. C. Hoffmann, M. de Groot, A. Hospers, 1996 The effect of the dust collection system on the flowpattern and separation efficiency of a gas cyclone, Canadian journal of chemical engineering, 74 (4), 464 470 .
29. S. Hogg, M. A. Leschziner, 1989 Computational of highly swirling confined flow with a Reynolds stress turbulence model, AIAA J,1(27), 57-63.
30. D. L. a. L. D. Iozia, 1989 Effect of cyclone dimensions on gas flow pattern and collection efficiency. Aerosol Science tech., 10 491 500 .
31. S. Jakirlic, K. Hanjalic, 2002 Modelling rotating and swirling turbulent flows: a perpetual challenge. AIAA J, 40 1984 96 .
32. Z. Ji, Z. Xiong, H. Chen, H. Wu, 2009 Experimental investigations on a cyclone separator performance at an extremely low particle concentration. Powder Tech. (191), 254 259 .
33. J. L. Jones, J. M. Arnold, C. A. Youngdahl, 1979 Erosion rates and patterns of the gas pilot plant's effluent cyclone.

34. F. Kaya, I. Karagoz, 2009 Numerical Investigation of performance characteristics of a cyclone prolonged with a dipleg. Che. eng. Journal, 151 39 45 .
35. J. Kim, 1990 Experimental study of particle collection by small cyclones, Aerosol Science tech., 12 1003 1015 .C. E. Lapple, 1951 Processes use many collector types, Chemical engineering, 58 144 151 .
36. D. Leith, W. Licht, 1972 Collection efficiency of cyclone type particle collector: A new theoretical approach, A.I.Ch.E Symposium series (Air-1971), 68, 196-206.
37. L. Ma, D. B. Ingham, X. Wen, 2000 Modelling of the fluid and particle penetration through small sampling cyclones. J Aerosol Sci (31), 1097-1119.
38. H. F. Meier, M. Mori, 1998 Gas-solid flow in cyclones: theEulerian-Eulerian approach. ComputChemEng, 22(Suppl 1):S641 S644 .
39. H. F. Meier, M. Mori, 1999 Anisotropic behavior of the Reynolds stress in gas and gas-solid flows in cyclones. Powder Tech. (101), 108 119 .
40. S. A. Morsi, A. J. Alexander, 1972 An investigation of particle trajectories in two phase flow systems. Fluid Mech. J., 55(2), 193-208.
41. E. Muschelknautz, 1972 Die Berechnung von Zyklonabscheidern fur Gase. Chem-Ing-Tech (44), 63-71.
42. M. Noppenberger, 2000 How to control erosion in FCC cyclone. World refining, 10(6), 36-38.
43. S. Obermair, 2003 Einfluss der Feststoffaustragsgeometrie auf die Abscheidung und den DruckverlusteinesGaszyklons, Technical University Graz.
44. K. Pant, C. T. Crowe, P. Irving, 2002 On the design of miniature cyclone for the collection of bioaerosols. Powder Tech., 125 260 265.
45. Parida, P. Chand, 1980 Turbulent swirl flow with gas-solid flow in cyclone. Che. Eng. Sci., 35(4), 949-954.
46. P. A. M. Patterson, R. J. Munz, 1989 Cyclone Collection Efficiencies at Very High Temperatures. Can. J. Chem. Eng, 37.
47. F. Qian, Z. Huang, G. Chen, M. Zhang, 2006 Numerical study of the separation characteristics in a cyclone of different inlet particle concentrations. Computers and chemical engineering, 31 1111 1122 .
48. Raoufi, M. Shams, H. Kanani, 2009 CFD analysis of flow field in square cyclones. Powder Tech. (191), 349 357 .

49. Saltzmann, 1984 Generalized performance characteristics of miniature cyclone for atmospheric particulate sampling. Am. Ind. Hyg. Assoc. J., 45 671 680 .
50. H. Shalaby, K. Pachler, K. Wozniak, G. Wozniak, 2005 Comparative study of the continuous phase flow in cyclone separator using different turbulence models. International J. of Numerical methods in fluids, 11(48), 1175 1198 -1197.
51. H. Shalaby, K. Wozniak, G. Wozniak, 2008 Numerical calculation of particle-laden cyclone separator flow using LES. Eng. app. of Comp. Fluid Mech., 2(4), 382-392.
52. B. Shepherd, C. E. Lapple, 1939 Flow pattern and pressure drop in cyclone dust collectors. Ind. Eng. Chem, 31 972 984 .
53. L. Shi, D. J. Bayless, G. Kremer, B. Stuart, 2006 CFD Simulation of the Influence of Temperature and Pressure on the Flow Pattern in Cyclones. Ind. Eng. Chem. Res. (45), 7667 7672.
54. M. S. K. Shin, D. S. Jang, J. D. Chung, M. Bohnet, 2005 Numerical and Experimental Study on a High Efficiency Cyclone Dust Separator for High Temperature and Pressurized Environments, Appl. Therm. Eng., (25), 1821 1835 .
55. M. D. Slack, R. O. Prasad, A. Bakker, F. Boysan, 2000 Advances in cyclones modeling using unstructured grids, Transactions of the Institution of Chemical engineers, 78A, 1098.
56. M. Sommerfeld, C. H. Ho, 2003 Numerical calculation of particle transport in turbulent wall bounded flows. Powder Tech., 131 1 6 .
57. Sproul 1970 Air pollution and its control, New York: Exposition Press.
58. Stairmand 1951 The design and performance of cyclone separators. Trans. Isntn. Chem. Engrs, 29 356 383 .
59. G. J. Sturgess, S. A. Syed, 1985 Calculation of a hollow-cone liquid spray in uniform airstream, J. Propul. Power (1)
60. P. Swift, 1969 Dust Control in Industry. Steam Heat. Engr., 38 453 456 .
61. H. J. Tengbergen, 1965 Comparative studies with cyclone, Staub, 25(11), 44-49.
62. J. Velilla, 2005 Study of the flow at a PFBC cyclonedipleg. University of Zaragoza.
63. G. Wan, G. Sun, X. Xue, M. Shi, 2008 Solids concentration simulation of different size particles in a cyclone separator, Powder Tech., 183 94 104 .

64. Wang, D. L. Xu, K. W. Chu, A. B. Yu, 2006 Numerical study of gas solid flow in a cyclone separator, Applied Mathematical Modeling, 30 1326 1342 .
65. S. Wang, M. Fang, Z. Luo, X. Li, M. Ni, K.. Chen, 1999 Instantaneous separation model of a square cyclone,. Powder Technology (102), 65 70 .
66. R. B. Xiang, K. W. Lee, 2005 Numerical study of flow field in cyclones of different height. Che. Eng. Sci., 44 877 883 .
67. A. Youngdahl, 1984 Nondestructive monitoring of erosive wear in transfer lines and cyclones at synfuels pilot plants, Paper presented at the Corrosion/84, International Corrosion Forum Devoted Exclusively to the Protection and Performance of Materials
68. S. Yuu, T. Jotaki, Y. Tomita, K. Yoshida, 1978 the reduction of pressure drop due to dust loading in a conventional cyclone. ChE. Eng. Sci., 33(12), 1573 1580.
69. Zhao, Y. Su, J. Zhang, 2006 Simulation of gas flow pattern and separation efficiency in cyclone with conventional single and spiral double inlet configuration. Chemical Engineering research and Design, A12 (84), 1158 1165.

CITATION

Utikar, R., Darmawan, N., Tade, M., Li, Q, Evans, G., Glenny, M. and Pareek, V. (2010). Hydrodynamic Simulation of Cyclone Separators, Computational Fluid Dynamics, Hyoung Woo Oh (Ed.), ISBN: 978-953-7619-59-6, InTech, DOI: 10.5772/7106.

CHAPTER 5

Numerical Simulation of the Flow Field in Circumferential Grooved Liquid Seals

Meng Zhang, Xiao-fang Wang, Sheng-li Xu, and Wei Wang

School of Energy and Power Engineering, Dalian University of Technology, No. 2, Linggong Road, Dalian, Liaoning 116024, China

ABSTRACT

Circumferential grooved liquid seals are utilized inside turbomachinery to provide non-contacting control of internal incompressible fluid leakage. Accurate prediction of the flow field is fundamental in producing robust and efficient designs. To validate the capabilities of the computational fluid dynamics FLUENT for incompressible fluid seal flow, comparisons of velocity parameters are made to the published experimental results and other CFD code for a circumferential grooved liquid seal. This work employs a pressure-based CFD code FLUENT to calculate the flow field in the seal, using four different turbulence models respectively. The velocity contours are compared with experimental values. It shows good overall agreement of the axial, radial, and azimuthal velocities in the through-flow jet, shear layer, and recirculation zone. Quantitative comparisons of velocity profiles at the center of the groove are made to experiment. This study verifies the prediction accuracy of three turbulence models. Various structures were considered to obtain a better understanding of the circumferential grooved liquid flow characteristics. The best groove structure to control leakage was also found within the limited designed seal. This study will provide a useful reference for designing the circumferential grooved liquid seal.

INTRODUCTION

Circumferential grooved liquid seals are used extensively in turbo-machineries to reduce leakage from high-pressure areas to low-pressure ones. When the leakage is reduced, there will be an accompanying increase in the efficiency of the device. Accurate prediction of the flow field is vital for reliable performance. However, grooved liquid seals have not been investigated thoroughly due to large pressure and velocity gradients near the wall, large recirculation, and high turbulence intensity present in the seals. To better investigate the complex leakage flow characteristics of circumferential grooved liquid seals, numerical simulated approaches have been developed and comparisons to experimental results for the flow field have not been performed by several researchers.

One of the first numerical CFD comparisons to experimental results for circumferential grooved liquid seals was made by Rhode et al. [1], who compared their prediction with the one-dimensional LDA measurements of Stoff [2] for swirl velocity and the measured pressure distributions of Morrison et al. [3]. Experimental data was limited to one plane per groove cavity and did not allow a thorough comparison to be made.

Nordmann et al. [4] calculated the leakage flow of circumferential grooved liquid seals based on Childs [5] finite length solution, which is derived from Hirs' [6] bulk-flow theory. Nordmann et al. modified Hirs' turbulent lubricant equations to account for the different friction factors in the circumferential and axial directions in the grooves.

Demko et al. [7] compared numerical predictions with experimental hotfilm measurements for the axial and swirl velocity components and turbulent kinetic energy in a grooved liquid annular seal. The results showed reasonable correlation of mean velocity both in the through-flow jet and in the recirculation zone.

Kilgore and Childs' experiment [8] provided an additional basis for evaluating the model for grooved liquid seals with different geometries and higher Reynolds numbers. The effects of the combined friction factor model could be seen in the predictions for leakage coefficients. The speed dependency of the friction factor model was particularly noticeable in the predictions for leakage coefficients. At lower pressure differential, leakage predictions and friction factor correlation were both poor. Better leakage predictions were obtained at higher pressure differential. Regardless of the difference in friction factor speed dependency, the predicted friction factors were closer to the measured results.

Iwatsubo et al. [9] developed a two-control-volume theory, featuring a description of the seal in land and groove sections, in which nonspeed dependent pressure loss coefficients inside the seal were introduced. Their theory gave qualitative agreement with experiment.

Marquette and Childs [10] provided a new theory, which was derived from Florjancic's [11] three-control-volume theory for circumferentially grooved liquid seals. Marquette featured the introduction of diverging flow in groove sections. Validation of the new analysis was achieved through a comparison with existing experimental data taken from Kilgore [12]. Leakage was reasonably well predicted. It provided all numerical parameters that had been set correctly. Each variation of a parameter yielded changes that always have a physical explanation. This showed that the physics of the phenomenon were well captured by three-control-volume theory.

Morrison and Robic [13] compared a commercial CFD code with Morrison's [14] three-dimensional LDA measurements for the whirling, circumferentially grooved liquid seals. The code did a good job predicting the pressure field for circumferentially grooved liquid seals and yielded similar results with LDA measurements. A more recent study by Jeffrey Moore and Palazzolo [15] employed a pressure-based CFD code SCISEAL developed by Athavale et al. [16] to calculate the flow field in a whirling, circumferentially grooved liquid seals. Comparisons of basic fluid parameters were made to Morrison's [14] three-dimensional LDA measurements for a spinning, circumferentially grooved seal. The results showed good qualitative agreement with experiment for the flow field.

Comparisons were also made between experimental results and predictions from CFD model by Childs et al. [17]. The model included the assumption that a groove was large enough to create separate lands within the seal, creating a zero or negligible pressure perturbation within the groove. Test results showed that even the deepest groove depth tested is not deep enough to satisfy this assumption.

Vijaykumar's [18] CFD simulations were performed on 2D grooved liquid seals to compare flows fields with experimental LDA data from Johnson [19] and to investigate the effects of preswirl, clearance, and Reynolds numbers. The three-dimensional velocity and kinetic energy dissipated were plotted as a function of the tooth with preceding cavity. Vijaykumar's CFD simulations did a fairly good job in predicting the flow field and the associated flow physics in a 2D grooved liquid seals seal.

The uncertainty of the turbulent model's accuracy for the estimate of the flow field may result in an inefficient seal design with a larger leakage flow disturbing the main flow at the inlet of the impeller of the pump company, and then the hydraulic efficiency decreases. In this work, predictions are made respectively using the RNG κ-ε turbulence, the standard κ-ε turbulence, the realizable κ-εturbulence, and the Shear-Stress Transport (SST) κ-ω turbulence models, to validate the capabilities of the computational fluid dynamics (CFD) for incompressible fluid seal flow and find a turbulence model with superior capability in predicting the incompressible flow field of the circumferential grooved seal over others. And it can be helpful in providing an accurate prediction of the flow field for a wide variety of operating conditions and flow parameters.

COMPUTATIONAL DESCRIPTIONS

Model

A two-dimensional CFD model is used in this study to calculate the steady incompressible flow within the seal. Water is employed as the sealing medium and the temperature is treated as constant value. The RNG κ-ε turbulence, the standard κ-ε turbulence, the realizable κ-ε turbulence, and the Shear-Stress Transport (SST) κ-ω turbulence models were employed to determine the leakage characteristics. And CFD analysis of the seal is performed by ANSYS FLUENT. A pressure based solver with finite volume method of discretization was used. A second order discretization scheme was used for the pressure, density, and momentum terms. First order upwind scheme was used for the turbulent kinetic energy and turbulent dissipation rate terms.

The turbulence kinetic energy, κ, and its rate of dissipation, ε, are obtained from the following transport equations:

$$\begin{aligned}\frac{\partial}{\partial t}(\rho k)+\frac{\partial}{\partial x_i}(\rho k u_i) &= \frac{\partial}{\partial x_j}\left[\left(\mu+\frac{\mu_t}{\sigma_k}\right)\frac{\partial k}{\partial x_j}\right] \\ &\quad + G_k + G_b - \rho\varepsilon - Y_M + S_k, \\ \frac{\partial}{\partial t}(\rho\varepsilon)+\frac{\partial}{\partial x_i}(\rho\varepsilon u_i) &= \frac{\partial}{\partial x_j}\left[\left(\mu+\frac{\mu_t}{\sigma_\varepsilon}\right)\frac{\partial \varepsilon}{\partial x_j}\right] \\ &\quad + C_{1\varepsilon}\frac{\varepsilon}{k}(G_k + C_{3\varepsilon}G_b) - C_{2\varepsilon}\rho\frac{\varepsilon^2}{k} + S_\varepsilon,\end{aligned} \tag{1}$$

where

$$\mu_t = \rho C_\mu \frac{k^2}{\varepsilon}, \tag{2}$$

$$G_k = -\rho u_i' u_j' \frac{\partial u_j}{\partial x_i}, \tag{3}$$

$$G_b = -\frac{1}{\rho}\left(\frac{\partial \rho}{\partial T}\right)_p g_i \frac{\mu_t}{0.85}\frac{\partial T}{\partial x_i}, \tag{4}$$

$$Y_M = 2\rho\varepsilon\frac{k}{\gamma RT}. \tag{5}$$

In these equations, G_k represents the generation of turbulence kinetic energy due to the mean velocity gradients, defined as (3); G_b is the generation of turbulence kinetic energy due to buoyancy, defined as (4); represents the contribution of the fluctuating dilatation in compressible turbulence to the overall dissipation rate, defined as (5). The model constants, Y_M, , and , $C_{1\varepsilon} = 1.44;\ C_{2\varepsilon} = 1.92;$ are the turbulent Prandtl numbers for κ and ε, respectively $C_\mu = 0.09;$ and $\sigma_k = 1.0,\ \sigma_\varepsilon = 1.3$ are user-defined source terms. The solution of two separate transport equations allows the turbulent velocity and length scales to be independently determined. It is a semiempirical model, and the derivation of the model equations relies on phenomenological considerations and empiricism [20].

The main difference among the κ-ε models is ε equation. The modeled transport equation for ε in the RNG κ-ε model is

$$\begin{aligned}\frac{\partial}{\partial t}(\rho\varepsilon) + \frac{\partial}{\partial x_i}(\rho\varepsilon u_i) &= \frac{\partial}{\partial x_j}\left(\alpha_\varepsilon \mu_{\text{eff}} \frac{\partial \varepsilon}{\partial x_j}\right) + C_{1\varepsilon}\frac{\varepsilon}{k} \\ &\quad \times (G_k + C_{3\varepsilon}G_b) - C_{2\varepsilon}\rho\frac{\varepsilon^2}{k} + S_\varepsilon - R_\varepsilon,\end{aligned} \tag{6}$$

where

$$R_\varepsilon = \frac{C_\mu \rho \eta^3 (1 - (\eta/4.38))}{1 + 0.012(Sk/\varepsilon)^3}\frac{\varepsilon^2}{k}. \tag{7}$$

One of the weaknesses of the standard κ-ε model lies with the modeled equation for the dissipation rate ε. The round-jet anomaly is considered to be mainly due to the modeled dissipation equation. The RNG model has an additional term in its ε equation that significantly improves the accuracy for rapidly strained flows. It yields a lower turbulent viscosity than the standard κ-ε model [20].

A new model equation for dissipation ε is based on the dynamic equation of the mean-square vorticity fluctuation. The modeled transport equation for ε in the realizable κ-ε model is

$$\begin{aligned}\frac{\partial}{\partial t}(\rho\varepsilon) + \frac{\partial}{\partial x_i}(\rho\varepsilon u_i) &= \frac{\partial}{\partial x_j}\left[\left(\mu + \frac{\mu_t}{\sigma_\varepsilon}\right)\frac{\partial \varepsilon}{\partial x_j}\right] + \rho C_1 S\varepsilon \\ &\quad - \rho C_2 \frac{\varepsilon^2}{k + \sqrt{\nu\varepsilon}} + C_{1\varepsilon}\frac{\varepsilon}{k}C_{3\varepsilon}G_b + S_\varepsilon.\end{aligned} \tag{8}$$

The production term in the ε equation of realizable κ-ε model does not involve the production of κ. The present form better represents the spectral energy transfer. Another feature is that the destruction term (the next to last term on the right-hand side of (8)) does not have any singularity. Its denominator never vanishes, even if κ vanishes. The benefit of the realizable κ-ε model is that it more accurately predicts the spreading rate of round jets. It is likely to provide superior performance for flows involving rotation, boundary layers under strong adverse pressure gradients, separation, and recirculation [20].

The Shear-Stress Transport (SST) κ-ω turbulence model was developed by Menter [21] to effectively blend the robust and accurate formulation of the κ-ω model in the near-wall region with the free-stream independence of the κ-ε model in the far field. To achieve this, the κ-ε model is converted into a κ-ω formulation. The definition of the turbulent viscosity is modified to account for the transport of the turbulent shear stress [20].

Dimensions and Conditions

The geometry of the seal is based on the Morrison's [14] experimental model and the simulation uses quad cells to create the entire grid. Figure 1 shows that the geometry consists of a labyrinth seal, a supply nozzle upstream of the seal portion, and the downstream region with increased clearance. The section of the supply expansible nozzle provides an annular flow path leading up to the seal entrance. There is a 0.508 mm gap between the inlet contour assembly and the rotor. The seal portion is sufficiently long to ensure that the flow is fully developed. The flow leaves through the region of large clearance. The geometry of the model will consist of the following dimensions: labyrinth seal length mm, rotor radius mm, and clearance mm.

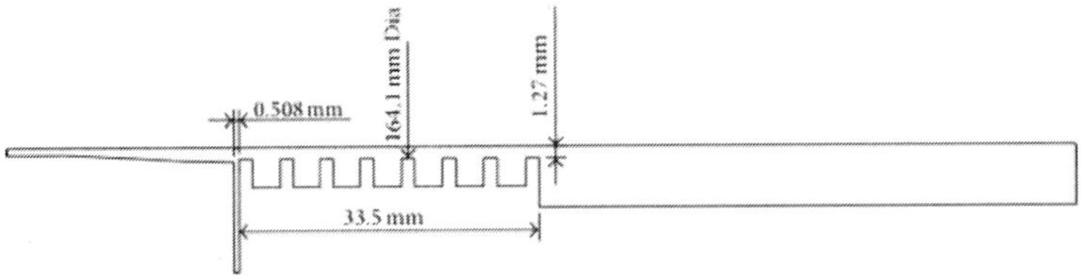

Figure 1: Seal geometry.

The geometry of the seal is considered axisymmetric and periodic boundary conditions are assumed in the circumferential direction. Mass flow inlet boundary is maintained at the leakage rate of 0.00486 m^3/s, which corresponds to a Reynolds number of 24,000. The flow direction will be normal to the boundary, and there will be no swirl component at the boundary for axisymmetric swirl. Outlet boundary condition is imposed pressure at 413.7 kPa. The rotor was imparted shaft rotation by specifying the rpm at 3600. The turbulence intensity at the core of a fully developed flow can be estimated from the following formula:

$$I = \frac{u'}{u_{avg}} = 0.16\,(Re\,D)^{0.125}, \tag{9}$$

Mesh

The 2D mesh applying the structured cells for the land and groove is created using a commercial mesh generator Gambit as shown in Figure 2. Because the change of the flow near the walls may be large, the mesh distribution was not uniform with a dense mesh in the annular region and a course mesh in the interior of the cavity. When generating a mesh, care is taken to maintain a smaller value, resulting in improved wall shear stress and pressure prediction.

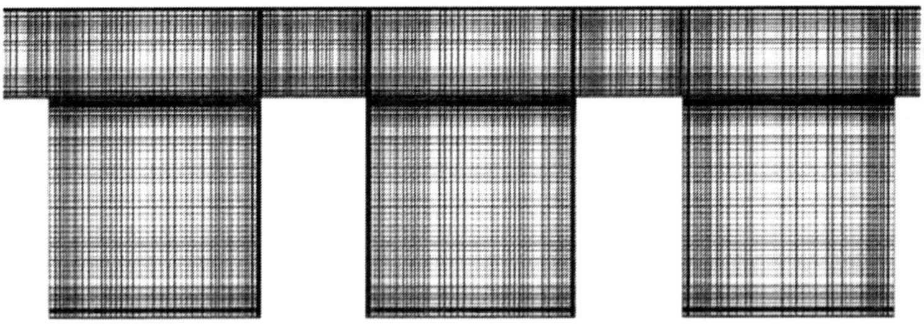

Figure 2: Labyrinth seal mesh.

Laws of the wall formulations model the sharp pressure gradients near the wall and are used with standard κ-ε turbulence model:

$$U^* = y^* \quad \text{for } y^* < 11.225$$
$$U^* = \frac{1}{\kappa}\ln\left(Ey^*\right) \quad \text{for } y^* > 11.225, \tag{10}$$

where

$$U^* \equiv \frac{U_P C_\mu^{1/4} \kappa_P^{1/2}}{\tau_w / \rho} \tag{11}$$

is the dimensionless velocity,

$$y^* \equiv \frac{\rho C_\mu^{1/4} \kappa_P^{1/2} y_P}{\mu} \tag{12}$$

is the dimensionless distance from the wall, and $C\mu$ = turbulent viscosity constant (= 0.09), κ = von Kármán constant (= 0.4187), E = empirical constant (= 9.793), UP = mean velocity of the fluid at the near-wall node P, κP = turbulence kinetic energy at the near-wall node P, yP = distance frompoint P to the wall, and μ = dynamic viscosity of the fluid. Thelaws-of-the-wall for mean velocity and pressure are based on thewall unit, $y*$, rather than$y$ + ($\equiv \rho u \tau y/\mu$). These quantities are approximately equal in equilibrium turbulent boundary layers [20].

Thewall$y*$ values are ensured less than 10 throughout the seal. And amesh independence density study is performed to investigate the effect of mesh density and to determine how fine of a mesh is required to capture the important velocity and pressure physics. The mesh is refined in the radial and axial direction dependently.

Figure 3 shows the effect of refining the mesh in the clearance and groove regions on the inlet average pressure,which is a representation for pressure loss of the seal. It is seen that increasing the number of nodes from 37000 to 66000 in the seal cross-section led to little change in the inlet average pressure showing that grid independence is established. To obtain a better velocity prediction, maximum density mesh will be used in subsequent calculations.

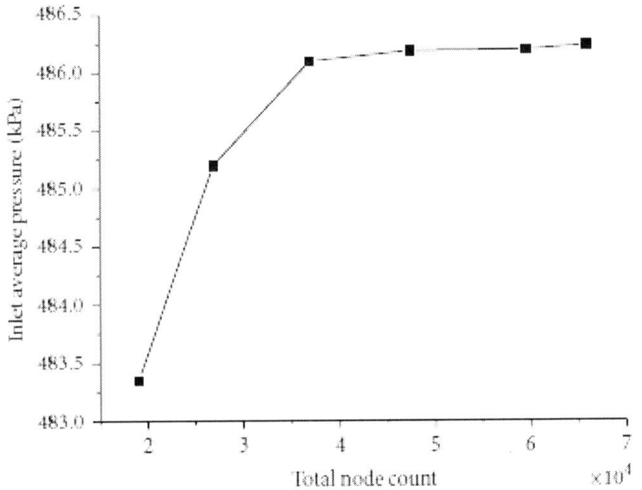

Figure 3: Mesh density study.

RESULTS AND DISCUSSION

Comparisons of Velocity Amplitude Contour Plots

Figure 4 shows the calculated axial velocity contours of the whole grooved seal. The pattern of vectors at each groove is similar. Figure 5 shows the representative velocity field for groove cross-section for the third seal cavity. The high through-flow jet generated due to a vena contracta effect from the groove to the clearance, which passes through the rotor tooth, remained parallel to and attached to the stator wall. When passing the groove, a smaller part of the core flow impinges obliquely on the front side of the downstream tooth, redirected towards the bottom edge of the rotor groove. And the jet current drives a single large vortex inside the cavity.

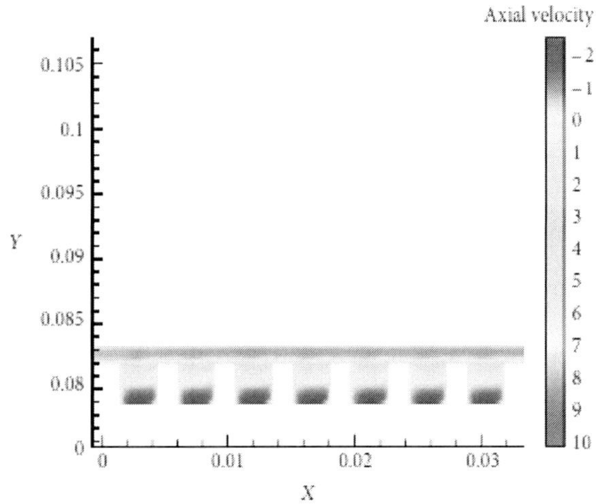

Figure 4: Axial velocity contours of the grooved seal.

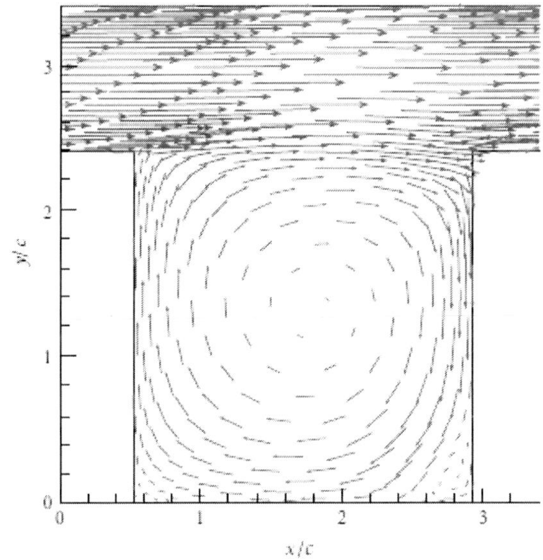

Figure 5: Velocity vector field for groove cross-section.

For the leakage rate of 0.00486 m3/s, the leakage Reynolds number was 24000, with an average axial mean velocity of 7.4 m/s in the clearance. The velocity contour plots are normalized with the average axial velocity (m/s). The coordinate represents the radial position () and the axial position () nondimensionalized with the clearance. Mean axial velocity contours for the third seal cavity are compared to the experimental plots [14] in Figure 6. The mean velocity contours of other seal cavity are essentially identical to those of the third cavity for the distributions of the velocity components. The results show good qualitative agreement in both the through-flow jet and the recirculation zone. The simulated field has similar contour lines in the cavity and the clearance as experimental results. The computed jet current core is observable with axial velocities exceeding the average leakage velocity by 15 percent as the same peak magnitude as the experimental results but slightly underpredicts the radial gradient of the axial velocity in the shear layer above the recirculation zone. Both of the two contour plots show a vena contracta effect with an increase in the radial gradient of the axial velocity as the fluid is accelerated from the large cavity into the small clearance above the downstream tooth and a rapid decrease gradient as the upstream vena contracta effect dissipates widely above the upstream tooth. Both the predicted and experimental jets have the same shear layer width with a rapid increase in the axial velocity near the stator surface.

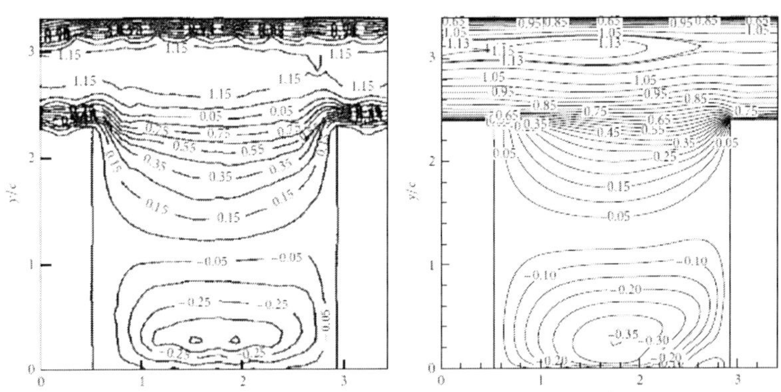

Figure 6: Axial velocity contour plots.

The predicted recirculation zone has essentially identical flow pattern as the experimental results. Inside the groove, the axial velocities are almost zero near the vertical walls of the rotor teeth and along a horizontal line near the center of the groove. In the top half of the recirculation zone, the computation has the same as the positive value contour lines of measurement but slightly underpredicts the radial gradient of the axial velocity in the shear layer. In the bottom half of the recirculation zone, the computation has negative contour lines; it differs from measurement by a reduction in the magnitude of the maximum negative axial velocity value.

Again, the radial velocity is normalized by the average axial velocity as shown in Figure 7. The predicted plots indicate that there is no observable radial velocity contour line above the teeth till the region near the top of the front side of the downstream tooth impinged by a smaller part of the core flow. Comparatively the experimental plots [14] are evident for the radial velocity contours with a significant radial velocity component present above the recirculation zone.

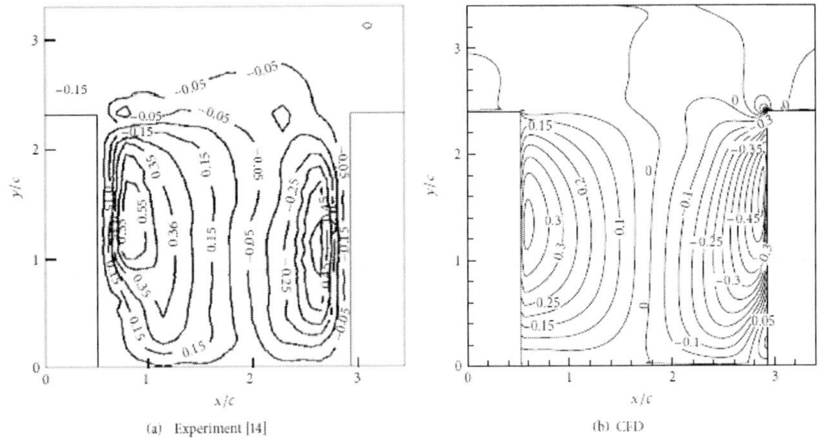

Figure 7: Radial velocity contour plots.

The predicted recirculation zone has essentially identical radial velocity contour shape as the experimental results. A smaller part of the core flow impinges the top of the front side of the downstream tooth, redirected to radial direction. This corresponds with the larger axial gradient of the radial velocity in the right part of the recirculation zone compared to the left part. In the left part of the recirculation zone, the computation has the positive value contour lines but slightly underpredicts the axial gradient of the radial velocity in the shear layer. The magnitude of the maximum positive radial velocity is much lower in the predicted plots () than in the experimental plots (). In the right part of the recirculation zone, the computation has negative contour lines, which differs from the measurement by a reduction in the magnitude of the maximum negative axial velocity value from to . Additionally the core position of the negative radial velocity contour is nearer to the downstream tooth, slightly deviating from the experiment results. This corresponds with the higher position of the tooth impinged by the small jet current compared to the experimental plots. It indicates that the model does not predict well the divergence angle of jet.

The velocity is normalized by the average azimuthal velocity (tooth tip surface velocity, m/s) as shown in Figure 8. Both the predicted and experimental cavities [14] have the same magnitude of the azimuthal velocity value () in the central region. The predicted plots have narrower shear layer width near the rotor wall than experiment. The standard near-wall formulations slightly underpredict the turbulence intensity in the shear layer of the groove and model the sharper gradients of azimuthal velocity. Also the azimuthal velocities above the teeth are underpredicted.

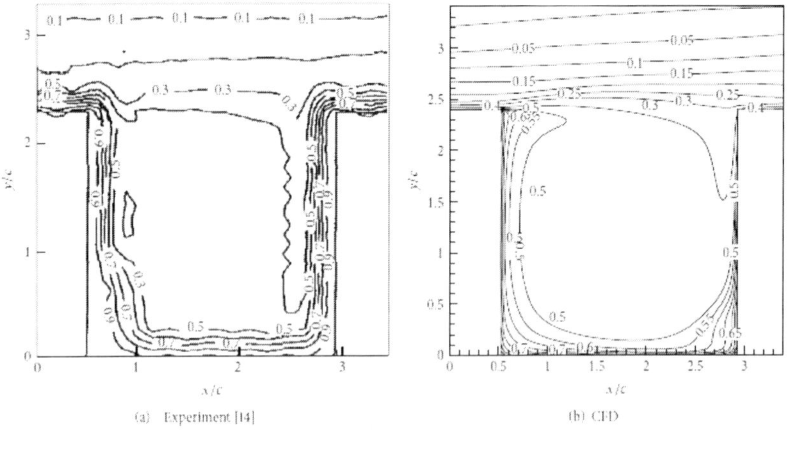

Figure 8: Azimuthal velocity contour plots ().

The κ-ε model velocity contours are compared with experimental values, showing good general agreement but differences near the cavity wall. It indicates that the model does not perform well in the weak shear layer in the high Reynolds number, because the κ-ε model is based on Boussinesq's isotropic eddy viscosity assumption, which has weakness in treatment of jet impingement and boundary layers.

Comparisons of Velocity Profiles

To validate the capabilities of the different turbulence models, quantitative comparisons of velocity profiles at sections in the cavities are made to experiment [14]. Figures 9 and 10 plot the axial and azimuthal velocity profiles at the center of the groove. The coordinate represents the radial position (Y) nondimensionalized with the clearance, while the abscissa for axial velocity is nondimensionalized with the average axial velocity (m/s). Figure 9 illustrates the axial velocity distribution of the seventh cavity. With the exception of the RNG κ-ε turbulence, the predicted velocity profiles using the standard κ-ε turbulence, the realizable κ-ε turbulence, the SST κ-ω turbulence, and SCISEAL [15] are essentially identical as the experiment. It validates the capability to predict the axial velocity of the incompressible flow. The CFD result predicts a steeper boundary layer near the stator wall and a through-flow jet with an axial velocity plateau. The RNG κ-ε turbulence and the standard κ-ε turbulence models yield identical peak velocities as the experiment. But another model slightly underpredicts the peak velocities. Various CFD models have a closely predicted through-flow jet as the experiment. It indicates their good capability to calculate the high Reynolds flow.

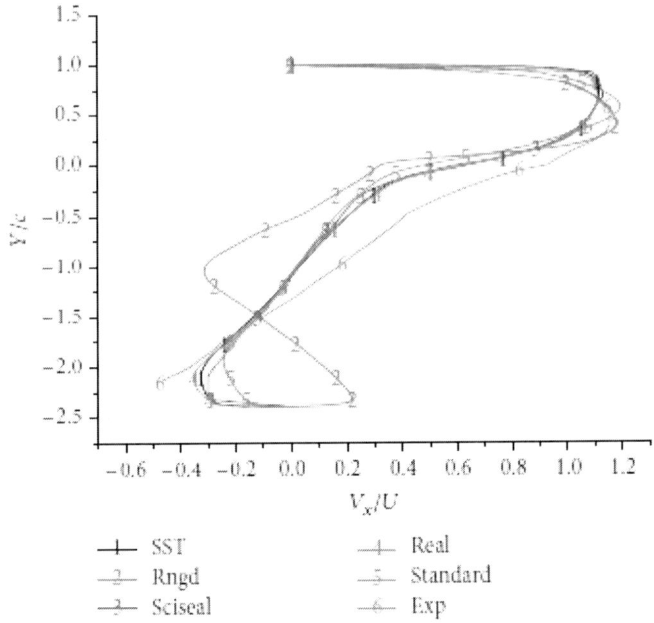

Figure 9: Axial velocity profiles (1, Shear-Stress Transport (SST) κ-ω turbulence model; 2, RNG κ-ε turbulence model; 3, CFD code SCISEAL [15]; 4, realizable κ-ε turbulence model; 5, standard κ-ε turbulence model; 6, experiment results [14]).

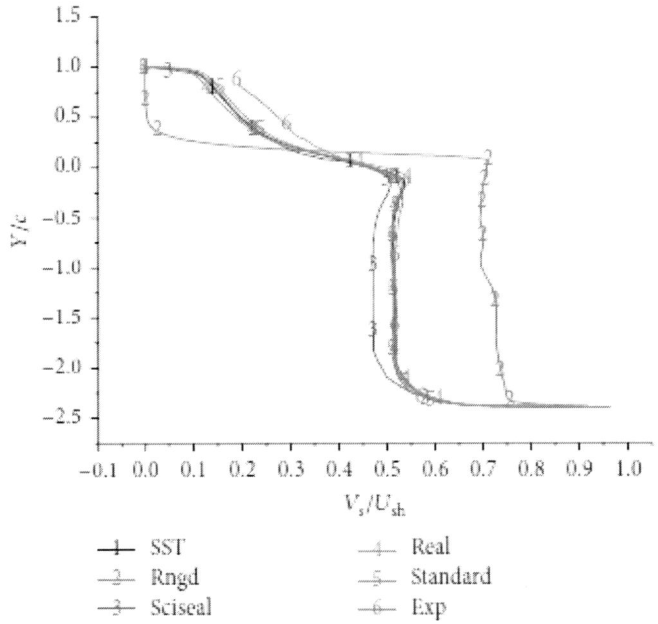

Figure 10: Azimuthal velocity profiles (1, Shear-Stress Transport (SST) κ-ω turbulence model; 2, RNG κ-ε turbulence model; 3, CFD code SCISEAL [15]; 4, realizable κ-ε turbulence model; 5, standard κ-ε turbulence model; 6, experiment results [14]).

In the groove, the standard κ-ε turbulence, the realizable κ-ε turbulence, the Shear-Stress Transport (SST) κ-ω turbulence model, and SCISEAL [15], predict essentially identical recirculation velocities as the experiment. But the RNG κ-ε turbulence predicts extremely different flow pattern with two vortexes. It is evident that the prediction of the recirculation in the groove is unexpectedly poor and shows no advantage for the vortexes. Comparatively other models have higher accuracy of the predicted vortex, with positive velocity in the top half of the groove and negative velocity in the bottom half just appearing to slightly underpredict the recirculation strength.

Near the bottom wall of the rotor groove, the standard near-wall formulations of the standard κ-ε turbulence slightly under-predict the turbulence intensity in the shear layer. It is the standard κ-ε turbulence's weakness in treatment of boundary layers. The realizable κ-ε model is likely to provide superior performance for flows involving rotation, boundary layers under strong adverse pressure gradients, and recirculation, because a new transport equation for the dissipation rate has been derived from an exact equation for the transport of the mean-square vorticity fluctuation. The Shear-Stress Transport (SST) κ-ω model shows advantage of treating boundary layers because the model effectively blends the robust and accurate formulation of the κ-ω model in the near-wall region. The definition of the turbulent viscosity is modified to account for the transport of the turbulent shear stress [20]. SCISEAL employs the two-layer model for a modification to the turbulence diffusion near the wall (inner layer) using an algebraic expression, while the turbulence kinetic energy equation is applied in both the inner and outer layers [15]. And as a consequence, these models' predictions of the axial velocities in boundary layers are expectedly accurate.

Figure 10 plots the azimuthal velocity profiles at the center of the groove. It was evident that the azimuthal velocity of the standard κ-ε turbulence, the realizable κ-ε turbulence, and the Shear-Stress Transport (SST) κ-ω turbulence model results accord well with the experiment, especially in the recirculation region. It indicates that these models could be used for high accurate azimuthal velocity quantitative analysis, which is fundamental for good rotordynamic characteristics assessment of grooved seals. SCISEAL [15] curve slightly deviates from the experiment results, under-predicting the azimuthal velocity in the groove and above the teeth. Therefore the standard κ-ε turbulence, the realizable κ-ε turbulence, and the Shear-Stress Transport (SST) κ-ω turbulence model provide higher accurate prediction of the azimuthal velocity than SCISEAL [15]. As can be observed from Figure 10, the RNG κ-ε turbulence has no advantage of the azimuthal velocity, with department forming the experiment in both the recirculation and the through-flow jet. The model grossly overpredicts the azimuthal velocity in the groove but severely underpredicts above the teeth.

Optimization of Groove Structure of Liquid Seals

This research attempts to validate the accuracy of different turbulence models for predicting the incompressible fluid seal flow.

Such predictive ability would greatly aid in the design of grooved seals which are critical pump components. The influence of the groove structure on the resistance of grooved liquid seals in pump is investigated. The friction factor is used for representing the resistance through the seals, which is defined as follows:

$$\lambda = -\frac{dP}{dx} \cdot \frac{D}{2\rho \overline{U}^2}, \qquad (13)$$

where$P$ is the static pressure, x is the axial seal coordinate, D is the hydraulic diameter, ρ is the density, and $U$is the average axial velocity.

The friction factors of sixteen different structure seals are predicted over a range of groove widths (Width/1.524mm= 1, 1.5, 2, 2.5) and depths (Depth/1.016mm = 1, 2, 3, 4) at the same clearance as the grooved seal (Depth/1.016mm = 3, Width/1.524mm = 2) in [14]. Figure 11 illustrates friction factor versus groove width and depth. As can be seen, the grooved seal (Depth/1.016mm= 1,Width/1.524mm= 2.5) has a higher resistance to leakage than other seals.

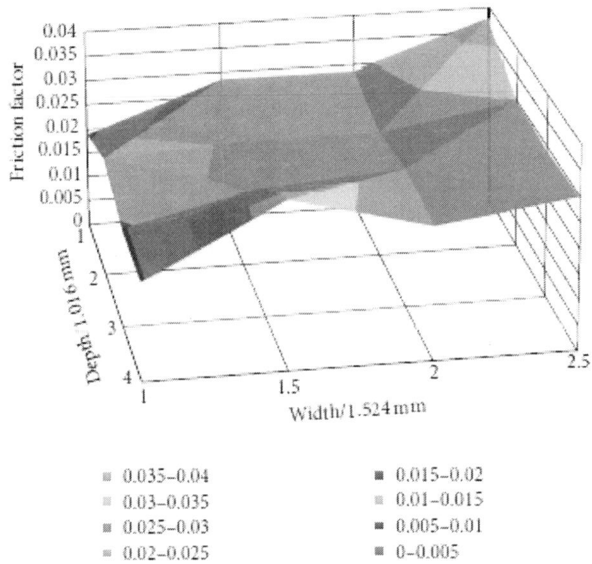

Figure 11: Friction factor versus groove width and depth.

The vector plots of Figure 12 show two different representative flow patterns for comparison of high resistance characteristics, such as Depth/1.016 mm = 1, Width/1.524 mm = 2.5 and Depth/1.016 mm = 4, Width/1.524 mm = 1. As shown in Figure 12(a), there exist two eddies in the groove (Depth/1.016 mm = 4, Width/1.524 mm = 1). The high through-flow jet generates due to a vena contracta effect from the groove to the clearance. When passing the groove, a smaller part of the core flow impinges obliquely on the front side of the downstream tooth, redirected towards the bottom edge of the rotor groove. The groove is deep enough to accommodate two large eddies in the opposite direction. Such a large recirculation is beneficial to energy dissipation because the kinetic energy is altered to the thermal energy in the eddy. It is speculated that the higher resistance of deep groove (Depth/1.016 mm = 4, Width/1.524 mm = 1) than the one (Depth/1.016 mm = 3, Width/1.524 mm = 2) in [14] is related to the large recirculation zone with two eddies. The other flow pattern of energy dissipation is the high-speed flow impinging on the seal wall. It can be found in Figure 12(b) that the stagnation area exists along the upwind surface of the groove due to the impingement of the high through-flow jet. Then a prolate eddy is formed in the shallow groove. It is consistent with the finding from Figure 13 that pressure distribution of the groove (Depth/1.016 mm = 1, Width/1.524 mm = 2.5) is heterogeneous due to energy dissipation, especially in the stagnation region at the corner. It is considered that the highest resistance to leakage in sixteen different structure seals as illustrated in Figure 11 is benefit from the impingement of the high through-flow jet. Thus the seal (Depth/1.016 mm = 1, Width/1.524 mm = 2.5) has the best performance of enhancing energy dissipation within the limited designed groove. The shallow grooved seal is recommended to be used in the pump in engineering.

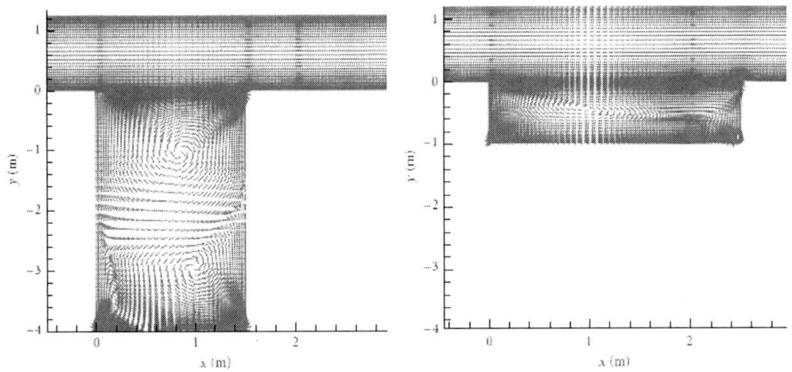

Figure 12: Velocity vector field for groove cross-section.

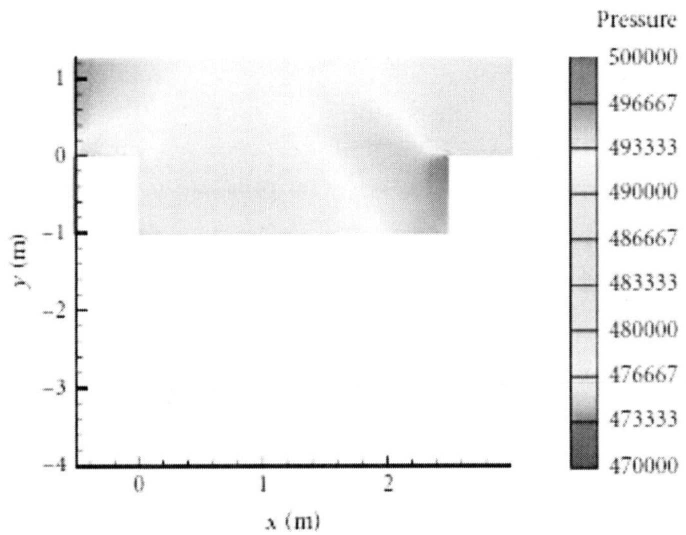

Figure 13: Pressure contours of the groove (Depth/1.016 mm = 1, Width/1.524 mm = 2.5).

CONCLUSIONS

This research attempts to validate the prediction accuracy of incompressible fluid seal flow in different CFD models by solving the general Reynolds Averaged Navier-Stokes equations along with the standard κ-ε turbulence, the realizable κ-ε turbulence, the RNG κ-ε turbulence, and the Shear-Stress Transport (SST) κ-ω turbulence model. Comparisons of velocity parameters are made to three-dimensional laser Doppler anemometer (LDA) measurements and the CFD code SCISEAL [15] for a circumferential grooved liquid seal.

Computed velocity contours are compared with experimental values, showing good general agreement but differences in detail near the wall, while through-flow jet, vortex, rotation, and vena contracta effect are well captured by the model. It validates the capability of the standard κ-ε turbulence for incompressible fluid seal flow.

To compare the predicted accuracy of the different turbulence models, quantitative comparisons of velocity profiles at sections in the cavities are made to experiment. The standard κ-ε turbulence, the realizable κ-ε turbulence, the Shear-Stress Transport (SST) κ-ω turbulence model, and SCISEAL [15] demonstrate good correlation in the axial velocities but slightly underpredict the recirculation in the groove. The RNG κ-ε turbulence predicts extremely a different flow pattern with two vortexes. The prediction of the recirculation in the groove is unexpectedly poor.

The standard κ-ε turbulence, the realizable κ-ε turbulence, and the Shear-Stress Transport (SST) κ-ω turbulence model really give satisfactory results in terms of the azimuthal velocity. It indicates that these models could be used for high accurate quantitative analysis, which is fundamental for good rotordynamic characteristics assessment of grooved seals. These models give more accurate results, showing substantial improvements over SCISEAL [15] in the rotational flow.

This study verifies the prediction accuracy of the standard κ-ε turbulence, the realizable κ-ε turbulence, and the Shear-Stress Transport (SST) κ-ω model. These models demonstrate superior capability in predicting the incompressible flow field of the circumferential grooved seal over SCISEAL [15].

Various structures were considered to obtain a better understanding of the circumferential grooved liquid flow characteristics. The best groove structure to control leakage was also found within the limited designed seal. This study will provide a useful reference in designing the circumferential grooved liquid seal.

Nomenclatures

:	Axial Reynolds number
:	Friction factor
:	Static pressure (Pa)
:	Hydraulic diameter (m)
:	Labyrinth seal length (mm)
:	Rotor radius (mm)
:	Seal clearance (mm)
:	Axial velocity (m/s)
:	Average axial velocity (m/s)
:	Radial velocity (m/s)
:	Azimuthal velocity (m/s)
:	Average azimuthal velocity (m/s)
:	Axial position (m)
:	Radial position (m)
:	Density (kg/m3)
:	Friction velocity (m/s)
:	Dynamic viscosity (Pas)
:	Turbulent kinetic energy (m2/s2)
:	Turbulent dissipation rate (m2/s3)
:	Turbulent viscosity constant
:	Von Kármán constant

: Empirical constant

: Mean velocity of the fluid at the near-wall node P (m/s)

: Turbulence kinetic energy at the near-wall node P (m2/s2)

: Distance from point P to the wall (m)

: Generation of turbulence kinetic energy due to the mean velocity gradients

: Generation of turbulence kinetic energy due to buoyancy

: Contribution of the fluctuating dilatation in compressible turbulence to the overall dissipation rate

Model constants for ε equations

: Turbulent Prandtl number for κ

: Turbulent Prandtl number for ε

: User-defined source term for κ

: User-defined source term for ε.

ACKNOWLEDGMENT

The authors appreciate the financial support from National Basic Research Program of China (973 Program).

REFERENCES

1. ANSYS Inc., ANSYS FLUENT 12.0 Theory Guide, 2009.
2. D. Childs, "Dynamic analysis of turbulent annular seals based on Hirs' lubrication equation," Journal of Lubrication Technology, vol. 105, no. 3, pp. 429–436, 1983.
3. D. L. Rhode, J. A. Demko, U. K. Traegner, G. L. Morrison, and S. R. Sobolik, "Prediction of incompressible flow in labyrinth seals," Journal of Fluids Engineering, vol. 108, no. 1, pp. 19–25, 1986.

4. D. W. Childs, M. Graviss, and L. E. Rodriguez, "Influence of groove size on the static and rotordynamic characteristics of short, laminar-flow annular seals," Journal of Tribology, vol. 129, no. 2, pp. 398–406, 2007.
5. F. R. Menter, "Two-equation eddy-viscosity turbulence models for engineering applications," AIAA Journal, vol. 32, no. 8, pp. 1598–1605, 1994.
6. G. G. Hirs, "A bulk-flow theory for turbulence in lubricant films," Journal of Lubrication Technology, vol. 95, no. 2, pp. 137–145, 1973.
7. G. L. Morrison and B. Robic, "Dynamic pressure measurements on the stator wall of whirling annular seals and numerical simulations," TRC Report TRC-SEAL-2-97, Texas A&M University, 1997.
8. G. L. Morrison, D. L. Rhode, K. C. Cogan, D. Chi, and J. A. Demko, "Labyrinth seals for incompressible flow," Final Report for NASA Contract NAS8-34536, 1983.
9. G. L. Morrison, M. C. Johnson, and G. B. Tatterson, "3-D laser anemometer measurements in a labyrinth seal," Journal of Engineering for Gas Turbines and Power, vol. 113, no. 1, pp. 119–125, 1991.
10. H. Stoff, "Incompressible flow in a labyrinth seal," Journal of Fluid Mechanics, vol. 100, no. 4, pp. 817–829, 1980.
11. J. A. Demko, G. L. Morrison, and D. L. Rhode, "The prediction and measurement of incompressible flow in a labyrinth seal," AIAA Paper 880190, 1988.
12. J. J. Kilgore and D. W. Childs, "Rotordynamic coefficients and leakage flow of circumferentially grooved liquid-seals," Journal of Fluids Engineering, vol. 112, no. 3, pp. 250–256, 1990.
13. J. J. Kilgore, Rotordynamic coefficients and leakage flow for parallel-grooved liquid-seals [M.S. thesis], A&M University, also available at Turhomachinery Laboratory, Report no. TL-SEAL-3-88, 1988.
14. J. Jeffrey Moore and A. B. Palazzolo, "CFD comparison to 3D laser anemometer and rotordynamic force measurements for grooved liquid annular seals," Journal of Tribology, vol. 121, no. 2, pp. 306–314, 1999.
15. M. C. Johnson, Development of a 3-D laser doppler anemometry system: with measurements in annular and labyrinth seals [Ph.D. dissertation], Texas A&M University, College Station, 1989.

16. M. M. Athavale, A. J. Przekwas, R. C. Hendricks, and A. Liang, "SCISEAL, A 3D CFD code for accurate analysis of fluid flow and forces in seals," in Proceedings of the Advanced ETO Propulsion Conference, May 1994.
17. O. R. Marquette and D. W. Childs, "An extended three-control-volume theory for circumferentially-grooved liquid seals," Journal of Tribology, vol. 118, no. 2, pp. 276–285, 1996. · ·
18. R. Nordmann, F. J. Dietzen, W. Janson, A. Frei, and S. Florjancic, "Rotordynamic coefficients and leakage flow of parallel grooved seals and smooth seals," in Rotordynamic Instability Problems in High-Performance, 1986.
19. S. Florjancic, Annular seals of high energy centrifugal pumps: a new theory and full scale measurement of rotordynamic coefficients and hydraulic friction factors [Ph.D. dissertation], Swiss Federal Institute of Technology, Zurich, Switzerland, 1990.
20. T. Iwatsubo, B. Sheng, and M. Ono, "Experiment of static and dynamic characteristics of spiral grooved seals, rotordynamic instability problems in high-performance turbomachinery," in Proceedings of a Workshop NASA CP, no. 3122, pp. 223–234, Texas A&M University, 1990.
21. Vijaykumar, Numerical Simulation of the flow field in 3D eccentric annular and 2D centered labyrinth seals for comparison with experimental LDA data [M.S. thesis], Texas A&M University, College Station, 2010.

CITATION

Meng Zhang, Xiao-fang Wang, Sheng-li Xu, and Wei Wang, "Numerical Simulation of the Flow Field in Circumferential Grooved Liquid Seals," Advances in Mechanical Engineering, vol. 2013, Article ID 797201, 10 pages, 2013. doi:10.1155/2013/797201

CHAPTER 6

Flow of a Generalised Newtonian Fluid Due to a Rotating Disk

P.T. Griffiths

School of Mathematics, University of Birmingham, Edgbaston, Birmingham B15 2TT, UK

ABSTRACT

The boundary-layer flow due to a rotating disk is considered for a number of generalised Newtonian fluid models. In the limit of large Reynolds number the flow inside the three-dimensional boundary-layer is determined via a similarity solution. Results for power-law and Bingham plastic fluids agree with previous investigations. We present solutions for fluids that adhere to the Carreau viscosity model. It is well known that unlike the power-law and Bingham models the Carreau model is applicable for vanishingly small, and infinitely large shear rates, as such we suggest these results provide a more accurate description of non-Newtonian rotating disk flow.

INTRODUCTION

The steady incompressible flow induced by the rotation of an infinite plane with uniform angular velocity is an exact solution of the Navier–Stokes equations, as was first described by von Kármán [1]. The flow is characterised by the lack of a radial pressure gradient near to the disk to balance the centrifugal forces so the fluid spirals outwards. The disk acts as a centrifugal fan, the fluid emanating from the disk being replaced by an axial flow directed back towards the surface of the disk.

Batchelor [2] showed that this type of flow is in fact just a limiting case of a whole number of flows with similarity solutions in which both the infinite plane and the fluid at infinity rotate with differing angular velocities. The corresponding limiting case, when the infinite plane is stationary and the fluid at infinity rotates at a constant angular velocity, was first described by Bödewadt [3].

A vast wealth of material exists concerning the solutions of the Newtonian rotating disk equations; the interested reader is referred to Zandbergen and Dijkstra [4]. The authors provide a thorough review of the major contributions made postdating von Kármán's seminal work.

Considerably less attention has been given to the corresponding non-Newtonian rotating disk problem. Mitschka [5] modified the von Kármán similarity solution to incorporate a power-law governing viscosity relationship. In this case the base flow is not an exact solution of the generalised Navier–Stokes equations and a boundary-layer approximation is required. Both Mitschka and Ulbrecht [6] and Andersson et al. [7] present basic flow solutions for shear-thickening and shear-thinning power-law fluids. However, the authors overlooked the importance of matching these boundary-layer solutions to an external flow. Denier and Hewitt [8] addressed this problem and presented corrected solutions for both cases, noting that the structure of the solutions is intrinsically different for shear-thickening and shear-thinning fluids.

More recently, Ahmadpour and Sadeghy [9] (subsequently referred to herein as AS) formally addressed the problem of the flow due to a rotating disk when one considers Bingham plastic fluids. Claiming to have found an exact solution to the problem, the authors are only able to present numerical solutions for specific values of the Reynolds number (Re) and dimensionless radius of the disk (r). Having not considered the boundary-layer formulation of the problem, the authors find that terms dependent on Re and r appear in the formulation of the governing base flow ODEs, and thus have the need to provide specific values for these constants during their numerical solution process.

In this study we determine steady mean flow solutions for power-law, Bingham and Carreau fluid models. The power-law results are essentially a review of the work of Denier and Hewitt [8] but are included here as they prove useful to compare with the results owing from the more complex Carreau model. By introducing the modified Bingham number used by Matsumoto et al. [10] in their film thickness investigation, we are able to determine a governing set of ODEs dependent solely on this parameter, these results are then compared to those of AS. Additionally, we present solutions for shear-thickening and shear-thinning Carreau fluids where now the flow is controlled by not one, but three dimensionless parameters. In Section 2 we formulate the problem in the general case, results are presented in Section 3 and are discussed in Section 4. Conclusions are drawn in Section 5.

FORMULATION

Consider the flow of a steady incompressible generalised Newtonian fluid due to a rotating disk located at $z^*=0$. The disk rotates about the z^*-axis with angular velocity Ω^*. Working in a reference frame that rotates with the disk, the continuity and Cauchy momentum equations are expressed as

$$\nabla \cdot \boldsymbol{u}^* = 0, \tag{1a}$$

$$\rho^*[\boldsymbol{u}^* \cdot \nabla \boldsymbol{u}^* + \boldsymbol{\Omega}^* \times (\boldsymbol{\Omega}^* \times \boldsymbol{r}^*) + 2\boldsymbol{\Omega}^* \times \boldsymbol{u}^*] = -\nabla p^* + \nabla \cdot \boldsymbol{\tau}^*. \tag{1b}$$

Here $\boldsymbol{u}^*=(u^*,v^*,w^*)$ are the velocity components in cylindrical polar coordinates(r^*,θ,z^*), the angular velocity vector is $\boldsymbol{\Omega}^*=(0,0,\Omega^*)$, the position vector is $\boldsymbol{r}^*=(r^*,0,z^*)$, the fluid density is ρ^* and p^* is the fluid pressure. The stress tensor $\boldsymbol{\tau}^*$for incompressible generalised Newtonian fluids is given by

$$\boldsymbol{\tau}^* = \mu^* \dot{\boldsymbol{\gamma}}^* \quad \text{with} \quad \mu^* = \mu^*(\dot{\gamma}^*), \tag{2}$$

where $\dot{\gamma}^* = \nabla \mathbf{u}^* + (\nabla \mathbf{u}^*)^T$ is the rate-of-strain tensor and $\mu^*(\dot{\gamma}^*)$ is the generalised Newtonian viscosity. The magnitude of the rate-of-strain tensor is $\dot{\gamma}^* = \sqrt{(\dot{\gamma}^* : \dot{\gamma}^*)/2}$. The governing relationships for $\mu^*(\dot{\gamma}^*)$ that will be considered herein are:

Power-law model $-\quad \mu^* = m^*(\dot{\gamma}^*)^{n-1}$, (3a)

Bingham model $-\quad \mu^* = \begin{cases} \infty & \text{for } \tau^* < \tau_y^*, \\ \mu_p^* + \tau_y^*(\dot{\gamma}^*)^{-1} & \text{for } \tau^* \geqslant \tau_y^*, \end{cases}$ (3b)

Carreau model $-\quad \mu^* = \mu_\infty^* + (\mu_0^* - \mu_\infty^*)[1 + (\lambda^*\dot{\gamma}^*)^2]^{(n-1)/2}$ (3c)

Here m* is the consistency coefficient and n is the fluid index, for n>1 the fluid is said to be shear-thickening, whilst for n<1 the fluid is said to be shear-thinning. The Newtonian viscosity relationship is recovered when $n = 1, \tau_y^* = 0$ and $\mu_0^* = \mu_\infty^*$, respectively. The plastic-shear-rate viscosityis μ_p^*, the magnitude of the shear stress tensor is $\tau^* = \sqrt{(\boldsymbol{\tau}^* : \boldsymbol{\tau}^*)/2}$ and τ_y^* is the yield stress. The infinite-shear-rate viscosityis μ_∞^*, the zero-shear-rate viscosity is μ_0^* and λ* is the characteristic time constant, often referred to as the 'relaxation time'.

Assuming the flow to be axisymmetric the components of the stress tensor are

$$\tau_{r^*r^*} = 2\mu^*\left(\frac{\partial u^*}{\partial r^*}\right), \tag{4a}$$

$$\tau_{\theta\theta} = 2\mu^*\left(\frac{u^*}{r^*}\right), \tag{4b}$$

$$\tau_{z^*z^*} = 2\mu^*\left(\frac{\partial w^*}{\partial z^*}\right), \tag{4c}$$

$$\tau_{r^*z^*} = \mu^*\left(\frac{\partial u^*}{\partial z^*} + \frac{\partial w^*}{\partial r^*}\right) = \tau_{z^*r^*}, \tag{4d}$$

$$\tau_{r^*\theta} = \mu^*\left[r^*\frac{\partial}{\partial r^*}\left(\frac{v^*}{r^*}\right)\right] = \tau_{\theta r^*}, \tag{4e}$$

$$\tau_{\theta z^*} = \mu^*\left(\frac{\partial v^*}{\partial z^*}\right) = \tau_{z^*\theta}, \tag{4f}$$

where the magnitude of the rate-of-strain tensor takes the form

$$\dot{\gamma}^* = \left\{ 2\left[\left(\frac{\partial u^*}{\partial r^*}\right)^2 + \left(\frac{u^*}{r^*}\right)^2 + \left(\frac{\partial w^*}{\partial z^*}\right)^2 \right] + \left[r^* \frac{\partial}{\partial r^*}\left(\frac{v^*}{r^*}\right) \right]^2 + \left(\frac{\partial v^*}{\partial z^*}\right)^2 + \left(\frac{\partial u^*}{\partial z^*} + \frac{\partial w^*}{\partial r^*}\right)^2 \right\}^{1/2}. \tag{5}$$

In the rotating frame of reference this system is closed subject to the boundary conditions

$$u^* = v^* = w^* = 0 \quad \text{at} \quad z^* = 0, \tag{6a}$$

$$u^* \to 0, \quad v^* \to -r^*\Omega^* \quad \text{as} \quad z^* \to \infty. \tag{6b}$$

We non-dimensionalise the system by writing

$$u(r,z) = \frac{u^*(r^*,z^*)}{l^*\Omega^*}, \quad v(r,z) = \frac{v^*(r^*,z^*)}{l^*\Omega^*}, \quad r = \frac{r^*}{l^*},$$

$$w(r,z) = \frac{w^*(r^*,z^*)}{\delta^* l^*\Omega^*}, \quad p(r,z) = \frac{p^*(r^*,z^*)}{\rho^*(l^*\Omega^*)^2}, \quad z = \frac{z^*}{\delta^* l^*}.$$

We note here that the axial coordinate and velocity component have been scaled by the boundary-layer thickness, δ*, this is in anticipation of a boundary-layer structure arising on the rotating disk. One finds that

$$\delta^* = Re^{-1/(q+1)} \quad \text{with} \quad Re = \frac{\rho^* \Omega^{*2-q} l^{*2}}{\sigma^*}, \tag{7}$$

where throughout the forthcoming analysis q=n for power-law fluids and q=1 for Bingham plastic and Carreau fluids, whilst $\sigma^* = m^*, \mu_p^*, \mu_\infty^*$ for power-law, Bingham plastic and Carreau fluids, respectively. Thus, the scaled governing equations are

$$\frac{1}{r}\frac{\partial(ru)}{\partial r} + \frac{\partial w}{\partial z} = 0, \tag{8a}$$

$$u\frac{\partial u}{\partial r} + w\frac{\partial u}{\partial z} - \frac{(v+r)^2}{r} = -\frac{\partial p}{\partial r} + \frac{\partial}{\partial z}\left(\mu\frac{\partial u}{\partial z}\right) + \frac{1}{Re^{2/(q+1)}}\left\{ \frac{2}{r}\frac{\partial}{\partial r}\left(\mu r\frac{\partial u}{\partial r}\right) + \frac{\partial}{\partial z}\left(\mu\frac{\partial w}{\partial r}\right) - \frac{2\mu u}{r^2} \right\}. \tag{8b}$$

$$u\frac{\partial v}{\partial r}+w\frac{\partial v}{\partial z}+\frac{uv}{r}+2u=\frac{\partial}{\partial z}\left(\mu\frac{\partial v}{\partial z}\right)+\frac{1}{Re^{2/(q+1)}}\left\{\frac{1}{r^2}\frac{\partial}{\partial r}\left[\mu r^3\frac{\partial}{\partial r}\left(\frac{v}{r}\right)\right]\right\}. \tag{8c}$$

$$u\frac{\partial w}{\partial r}+w\frac{\partial w}{\partial z}=-Re^{2/(q+1)}\frac{\partial p}{\partial z}+\frac{1}{r}\frac{\partial}{\partial r}\left(\mu r\frac{\partial u}{\partial z}\right)+2\frac{\partial}{\partial z}\left(\mu\frac{\partial w}{\partial z}\right)+\frac{1}{Re^{2/(q+1)}}\left\{\frac{1}{r}\frac{\partial}{\partial r}\left(\mu r\frac{\partial w}{\partial r}\right)\right\}. \tag{8d}$$

where the dimensionless viscosity functions (in the yielded region when considering Bingham plastic fluids) μ are defined as

Power-law model – $\mu=(\hat{\mu})^{n-1}$. (8e)

Bingham model – $\mu=1+2rB_n(\hat{\mu})^{-1}$. (8f)

Carreau model – $\mu=1+c_0[1+(k\hat{\mu}/r)^2]^{(n-1)/2}$. (8g)

where $B_n=\tau_y^*/\left(2r^*\Omega^*\sqrt{\mu_p^*\rho^*\Omega^*}\right)$ is the Bingham number defined by Matsumoto et al.[10]. The authors experimental investigations have shown this quantity to be O(1) for flows with $Re \gg 1$. The Carreau viscosity ratio is $c_0=(\mu_0^*-\mu_\infty^*)/\mu_\infty^*$ and $k=r^*\lambda^*\Omega^*\sqrt{\rho^*\Omega^*/\mu_\infty^*}$ is the dimensionless equivalent of the constant λ^*. Here

$$\hat{\mu}=\left[\left(\frac{\partial u}{\partial z}\right)^2+\left(\frac{\partial v}{\partial z}\right)^2+\mathcal{L}_{\hat{\mu}}\right]^{1/2}. \tag{8h}$$

the higher order terms, $\mathcal{L}_{\hat{\mu}}$, that contribute to the generalised viscosity, are given in theAppendix A for completeness. We note that the expressions for B_n and k can be simplified when one considers the modified Reynolds number, $R=rRe^{1/2}$, based on the boundary-layer thickness and the local azimuthal velocity of the disk. In that case

$$B_n=\frac{\tau_y^*}{2\mu_p^*\Omega^*R}, \tag{9a}$$

$$k=\lambda^*\Omega^*R. \tag{9b}$$

We are now able to make a boundary-layer approximation by eliminating terms involving inverse powers of the Reynolds number since here we assume that $Re \gg 1$. Inside the boundary-layer we assume a solution to (8a), (8b), (8c), (8d), (8e), (8f), (8g) and (8h)given that the velocity components, pressure and viscosity have the following asymptotic expansions

$$u(r,\tilde{z}) = u_0(r,z) + Re^{-2/(q+1)}u_1(r,z) + \cdots, \tag{10a}$$

$$v(r,\tilde{z}) = v_0(r,z) + Re^{-2/(q+1)}v_1(r,z) + \cdots, \tag{10b}$$

$$w(r,\tilde{z}) = w_0(r,z) + Re^{-2/(q+1)}w_1(r,z) + \cdots, \tag{10c}$$

$$p(r,\tilde{z}) = p_0(r,z) + Re^{-2/(q+1)}p_1(r,z) + \cdots, \tag{10d}$$

$$\hat{\mu}(r,\tilde{z}) = \hat{\mu}_0(r,z) + Re^{-2/(q+1)}\hat{\mu}_1(r,z) + \cdots, \tag{10e}$$

where $\tilde{z} = z^*/l^* = Re^{-1/(q+1)}z$ is the coordinate corresponding to the region outside of the boundary-layer. To leading order we see that $p_0=p_0(r)$, however, since it is assumed that inside the boundary-layer the pressure is a function of z only we have that $p_0\equiv 0$.

Having isolated the dominant viscous terms via the boundary-layer approximation, we arrive at the leading order equations that must be solved in order to determine the steady mean flow relative to the disk

$$\frac{1}{r}\frac{\partial(ru_0)}{\partial r} + \frac{\partial w_0}{\partial z} = 0, \tag{11a}$$

$$u_0\frac{\partial u_0}{\partial r} + w_0\frac{\partial u_0}{\partial z} - \frac{(v_0+r)^2}{r} = \frac{\partial}{\partial z}\left(\mu_0\frac{\partial u_0}{\partial z}\right), \tag{11b}$$

$$u_0\frac{\partial v_0}{\partial r} + w_0\frac{\partial v_0}{\partial z} + \frac{u_0 v_0}{r} + 2u_0 = \frac{\partial}{\partial z}\left(\mu_0\frac{\partial v_0}{\partial z}\right), \tag{11c}$$

$$u_0\frac{\partial w_0}{\partial r} + w_0\frac{\partial w_0}{\partial z} + \frac{\partial p_1}{\partial z} = \frac{1}{r}\frac{\partial}{\partial r}\left(\mu_0 r\frac{\partial u_0}{\partial z}\right) + 2\frac{\partial}{\partial z}\left(\mu_0\frac{\partial w_0}{\partial z}\right), \tag{11d}$$

where the viscosity functions are given by

$$\text{Power-law model} \quad - \quad \mu_0 = (\hat{\mu}_0)^{n-1}, \tag{11e}$$

$$\text{Bingham model} \quad - \quad \mu_0 = 1 + 2rB_n(\hat{\mu}_0)^{-1}, \tag{11f}$$

$$\text{Carreau model} \quad - \quad \mu_0 = 1 + c_0[1 + (k\hat{\mu}_0/r)^2]^{(n-1)/2}, \tag{11g}$$

with

$$\hat{\mu}_0 = \left[\left(\frac{\partial u_0}{\partial z}\right)^2 + \left(\frac{\partial v_0}{\partial z}\right)^2\right]^{1/2}. \tag{11h}$$

To solve for the basic flow inside the boundary-layer we introduce the modified von Kármán [1] similarity solution in the form

$$\boldsymbol{u}_0 = \left[r\bar{u}(\eta), r\bar{v}(\eta), r^{(q-1)/(q+1)}\bar{w}(\eta)\right], \quad p_1 = r^{2(q-1)/(q+1)}\bar{p}(\eta), \tag{12}$$

where $\eta = r^{(1-q)/(q+1)}z$, thus for Bingham plastic and Carreau fluids η reduces to z. Substituting this form for the similarity solution into (11a), (11b), (11c), (11d), (11e),(11f) and (11g) we find that $\bar{u}, \bar{v}, \bar{w}$ and $\bar{p}$ must satisfy

where the primes denote differentiation with respect to η and

$$-2\bar{u} + \frac{q-1}{q+1}\eta\bar{u}' = \bar{w}', \tag{13a}$$

$$\bar{u}^2 - (\bar{v}+1)^2 + \left(\bar{w} + \frac{1-q}{q+1}\eta\bar{u}\right)\bar{u}' = (\bar{\mu}\bar{u}')', \tag{13b}$$

$$2\bar{u}(\bar{v}+1) + \left(\bar{w} + \frac{1-q}{q+1}\eta\bar{u}\right)\bar{v}' = (\bar{\mu}\bar{v}')', \tag{13c}$$

$$\frac{q-1}{q+1}\bar{u}(\bar{w} - \eta\bar{w}') + \bar{w}\bar{w}' + \bar{p}' = \bar{v}\bar{u}' - 2\bar{\mu}'\bar{u} + (\bar{\mu}\bar{w}')', \tag{13d}$$

where the primes denote differentiation with respect to η and

Power-law model – $\bar{\mu} = (\bar{u}'^2 + \bar{v}'^2)^{(n-1)/2}$, (13e)

Bingham model – $\bar{\mu} = 1 + 2B_n(\bar{u}'^2 + \bar{v}'^2)^{-1/2}$, (13f)

Carreau model – $\bar{\mu} = 1 + c_0[1 + k^2(\bar{u}'^2 + \bar{v}'^2)]^{(n-1)/2}$ (13g)

The model dependent form of $\bar{v}$ is given in the Appendix B along with results for the pressure correction term, $\bar{p}$, which, as noted by Denier and Hewitt [8], is determined *a posteriori* once the velocity components have been calculated. By rearranging (13b) and (13c) we isolate terms involving $\bar{u}''$ and $\bar{v}''$; having such expressions allows us to formulate a system of five first order ordinary differential equations in five unknowns $\bar{u}, \bar{v}, \bar{u}', \bar{v}'$ and $\bar{w}$, that must be solved subject to

$$\bar{u} = \bar{v} = \bar{w} = 0 \quad \text{at} \quad \eta = 0, \tag{14a}$$

$$\bar{u} \to 0, \quad \bar{v} \to -1 \quad \text{as} \quad \eta \to \infty. \tag{14b}$$

Solving (13a), (13b), (13c), (13d), (13e), (13f) and (13g) subject to (14a) and (14b) requires a full numerical solution. We employ a shooting method that utilises a fourth-order Runge-Kutta quadrature routine to perform the numerical integration of the differential equations twinned with a Newton iteration scheme to determine the values of the unknowns $\bar{u}(0) = \bar{u}_0$ and $\bar{v}(0) = \bar{v}_0$.

STEADY MEAN FLOW

In much the same way as Denier and Hewitt [8], we firstly consider the asymptotic form of the functions $\bar{u}, \bar{v}$ and $\bar{w}$ as $\eta \to \infty$. In the limit as $\eta \to \infty$ we have that

$$\bar{w}_\infty \bar{u}' \sim (\bar{\mu}\bar{u}')', \tag{15a}$$

$$\bar{w}_\infty \bar{v}' \sim (\bar{\mu}\bar{v}')', \tag{15b}$$

Where

$$\bar{w}_\infty = -\frac{(3q+1)}{(q+1)} \int_0^\infty \bar{u}\, d\eta < 0.$$

For shear-thinning power-law fluids Denier and Hewitt [8] determined that $(\bar{u}, \bar{v}) \sim \eta^{n/(n-1)}$ as $\eta \to \infty$ as $\eta \to \infty$ and that $n/(n-1)<1$ to ensure w is bounded in the far-field. Thus, in order to produce boundary-layer solutions that match to an external flow we are only able to consider shear-thinning fluids with n in the range $0.5<n<1$. For shear-thickening fluids the analysis is somewhat more involved. Denier and Hewitt [8] showed that the solutions for $n>1$ become non-differentiable at a critical location $\eta=\eta_c$. However, it is noted that in this case the singularity can be completely regularised and thus the solutions can be matched to an external flow. We will not consider shear-thickening power-law fluids here, the interested reader is referred to Denier and Hewitt [8]for the full details regarding flows with n>1 and for a lengthy discussion on shear-thinning flows with 0<n<0.5,n=0.5 and 0.5<n<1.

For Bingham plastic fluids we find that $(\bar{u}, \bar{v}) \sim \exp(\bar{w}_\infty \eta)$ as $\eta \to \infty$, as is the case for Newtonian fluids, see Cochran [11]. This is particularly interesting to note as it indicates that the large η form of the solutions has no specific dependence on the Bingham number whatsoever. The most important consequence of this result is that the velocity solutions will remain bounded as $\eta \to \infty$ for all values of the Bingham number, provided, of course, that $\bar{w}_\infty \leqslant 0$.

Utilising the fact that $(\bar{u}', \bar{v}') \ll 1$ as $\eta \to \infty$ we find that $(\bar{u}, \bar{v}) \sim \exp[\bar{w}_\infty \eta/(1 + c_0)]$ for Carreau fluids as $\eta \to \infty$. We note here that the exponential decay of the velocity functions has no dependence on either *k* or the power-law index and, more importantly, the solutions are bounded for all shear-thinning values of *n*, whilst for n>1 no such 'finite width' crisis is encountered whereby the solutions become non-differentiable at a critical location, as was first noted by Dabrowski [12]. We reiterate that these solutions will exhibit exponential decay only when $\bar{w}_\infty \leqslant 0$. However, for the cases considered within this study, with c_0=1 and k=100,300,500, we find that $\bar{w}_\infty$ is in fact always negative for 0<n<1, whilst $\bar{w}_\infty$ is monotonically decreasing as n increases above unity. Hence these solutions are indeed always bounded as $\eta \to \infty$.

Having determined the large η form of the velocity functions we are then able to apply suitable asymptotic conditions as $\eta \to \infty$ within the numerical integration scheme, ensuring the correct far-field decay. This is achieved via the introduction of far-field mixed boundary conditions that relate $\bar{u}$ and $\bar{v}$ to their respective first derivatives. Solutions of (13a), (13b), (13c), (13d), (13e), (13f) and (13g) subject to (14a) and (14b) for a range of values of n, B_n and *k* are presented in Table 1 and Table 2 and Fig. 1, Fig. 4, Fig. 3 and Fig. 4, where the value of c_0 for Carreau fluids is held fixed at unity, thus making this study consistent with that of Dabrowski [12]. Where not stated, the value of 'infinity' is taken to be η_∞=20 in all cases, excluding the power-law analysis, this is sufficient to provide suitably converged results where the aforementioned asymptotic conditions are satisfied to within the desired tolerance of $\mathcal{O}(10^{-10})$.

Table 1: Numerical values of the basic flow parameters for Newtonian, power-law, Bingham plastic and Carreau fluids. The value of η_∞ employed for each power-law calculation is included, we note that these are not necessarily the same as used by Denier and Hewitt [8]. Solutions for n=1 are also included for Carreau fluids, these differ from the Newtonian results as in this case the viscosity function is effectively set to $\bar{\mu} = 1 + c_0$, rather than unity.

Newtonian fluids	$\bar{u}_0$	$-\bar{v}_0$	$-\bar{w}_\infty$	
$n - 1 = B_n = c_0 = 0$	0.5102	0.6159	0.8845	
Power-law fluids: $n =$	$\bar{u}_0$	$-\bar{v}_0$	η_∞	$-\bar{w}(\eta_\infty)$
0.9	0.5069	0.6243	50	0.9698
0.8	0.5039	0.6362	100	1.0957
0.7	0.5017	0.6532	200	1.3055
0.6	0.5005	0.6778	400	1.7209
Bingham plastic fluids: $B_n =$	$\bar{u}_0$	$-\bar{v}_0$	$-\bar{w}_\infty$	
0.1	0.4247	0.5900	0.7815	
0.2	0.3586	0.5685	0.7158	
0.3	0.3077	0.5489	0.6668	
0.4	0.2681	0.5308	0.6282	
0.5	0.2368	0.5141	0.5968	
Carreau fluids: $k = 100, n =$	$\bar{u}_0$	$-\bar{v}_0$	$-\bar{w}_\infty$	
0.25	0.5007	0.6026	0.9669	
0.5	0.4842	0.5789	1.0124	
0.75	0.4415	0.5250	1.0980	
0.95	0.3793	0.4554	1.2137	
1	0.3608	0.4355	1.2508	
1.05	0.3418	0.4153	1.2916	
1.25	0.2678	0.3368	1.4902	
1.5	0.1950	0.2562	1.8131	
1.75	0.1457	0.1986	2.2015	

Table 2: Numerical values of the basic flow parameters for shear-thinning and shear-thickening Carreau fluids for increasing values of k

Carreau fluids: $n = 0.5, k =$	$\bar{u}_0$	$-\bar{v}_0$	$-\bar{w}_\infty$
100	0.4842	0.5789	1.0124
300	0.4948	0.5940	0.9639
500	0.4982	0.5988	0.9473
Carreau fluids: $n = 1.5, k =$	$\bar{u}_0$	$-\bar{v}_0$	$-\bar{w}_\infty$
100	0.1950	0.2562	1.8131
300	0.1601	0.2111	2.1743
500	0.1456	0.1923	2.3776

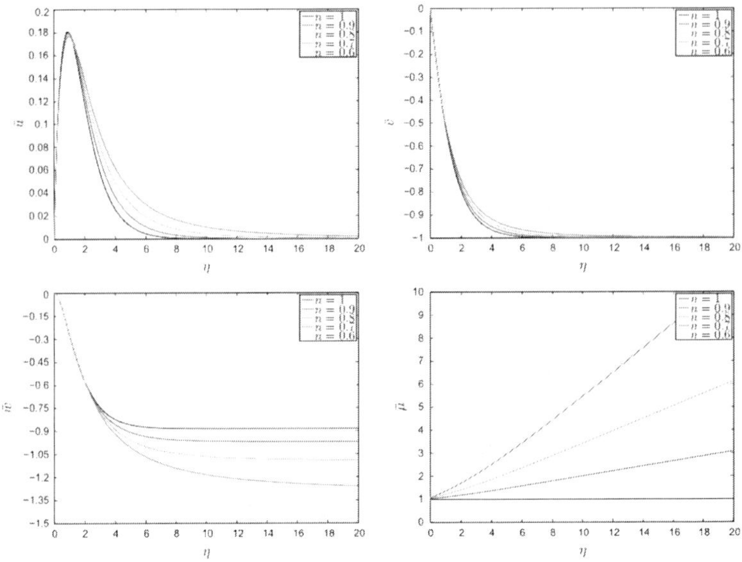

Figure 1: Plots of $\bar{u}, \bar{v}, \bar{w}$ and $\bar{\mu}$ versus η for power-law fluids with n=1,0.9,0.8,0.7,0.6.

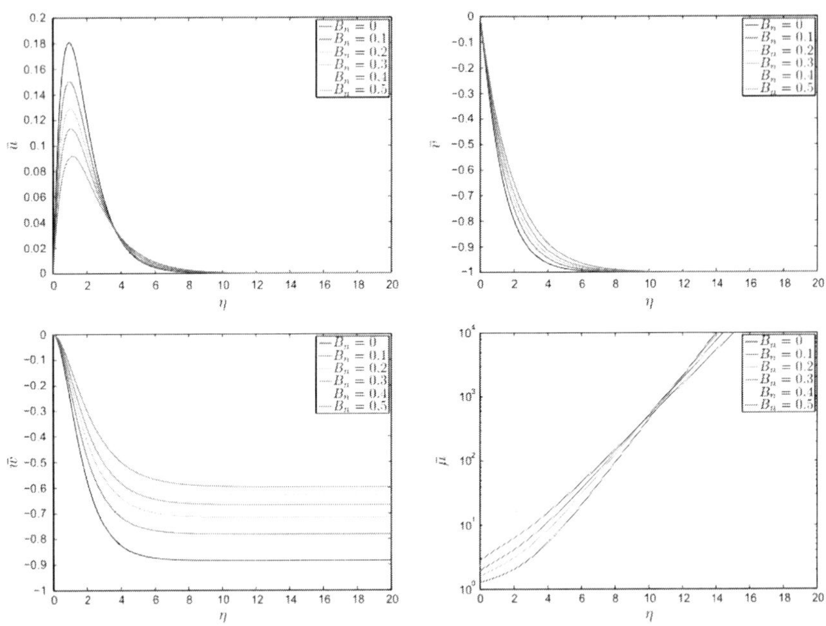

Figure 2: Plots of $\bar{u}, \bar{v}, \bar{w}$ and $\bar{\mu}$ versus η for Bingham plastic fluids with B_n=0,0.1,0.2,0.3,0.4,0.5.

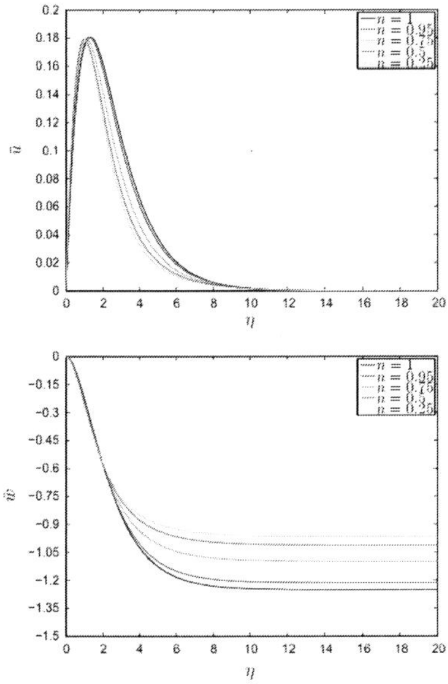

Figure 3: Plots of $\bar{u}, \bar{v}, \bar{w}$ and $\bar{\mu}$ versus η for Carreau fluids with n=1,0.95,0.75,0.5,0.25 and k=100.

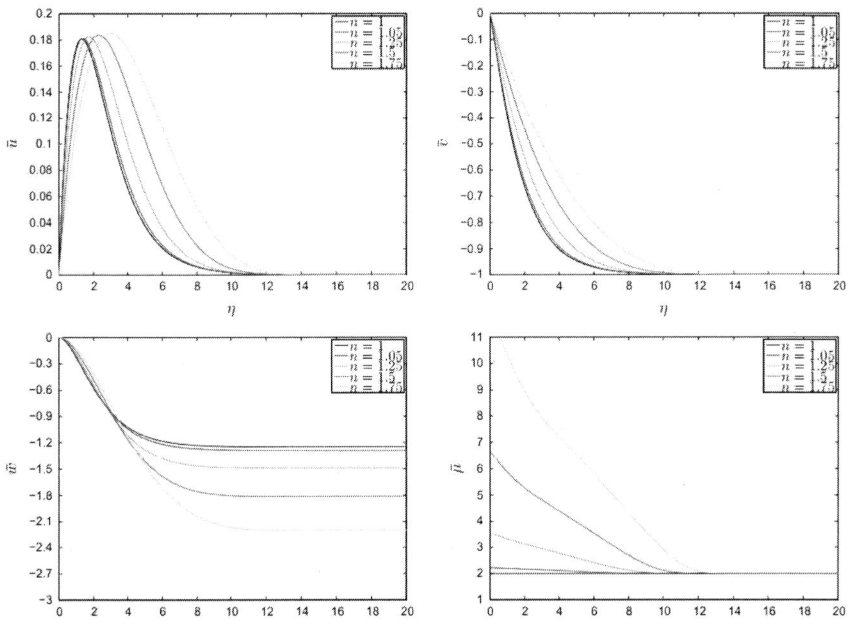

Figure 4: Plots of $\bar{u}, \bar{v}, \bar{w}$ and $\bar{\mu}$ versus η for Carreau fluids with n=1,1.05,1.25,1.5,1.75 and k=100.

DISCUSSION

The solutions presented in Fig. 1 are identical to those of Denier and Hewitt [8] and as such we forgo any detailed analysis, noting only that the value of η_∞ employed within the numerical scheme for n=0.7 and n=0.6 produces differing values of $-\bar{w}(\eta_\infty)$, for the respective values of n, when compared to the results of Denier and Hewitt [8]. This is as to be expected and is due to the ever slowing decay of the velocity functions as n approaches the critical limit of n=0.5. These results are included in this study for completeness whilst also serving as an aid to compare shear-thinning results owing from the Carreau model. Plots of the viscosity function, $\bar{\mu}$, are also included in Fig. 1; the growth of these functions appears to be linear with respect to the similarity coordinate. This observation is easily derived from the asymptotic analysis contained within the previous section. Given that $\bar{u} = A\eta^{n/(n-1)} + \cdots$ and $\bar{v} = B\eta^{n/(n-1)} + \cdots$ as $\eta\to\infty$, where A and B are constants, we find that $\bar{\mu} = C\eta + \cdots$ as $\eta\to\infty$, where $C=[n^2(A^2+B^2)/(n-1)^2](n-1)/^2$. From the governing non-linear ODEs (13a), (13b),(13c), (13d), (13e), (13f) and (13g) it is in fact possible to show that $\bar{\mu}' \to (n-1)\bar{w}_\infty/n$ as $\eta\to\infty$. Thus for 0.5<n<1 the viscosity function remains unbounded inside the boundary-layer and instead tends to a constant gradient as $\eta\to\infty$. Clearly this behaviour is unphysical, predicting that fluid is entrained into the boundary-layer with unbounded viscosity as the axial distance is increased, this being a result of the failure of the power-law model to accurately model shear-thinning flows in the limit as $\dot{\gamma}^* \to 0$. Analogously, one would find that $\bar{\mu} \to 0$ as $\eta \to \eta_c$ for shear-thickening power-law fluids for the same reason.

The results for Bingham plastic fluids are presented in Fig. 2. We observe a significant reduction in the peak of the cross-flow velocity component, $\bar{u}$, as B_n increases from zero, whilst the component of the azimuthal velocity, $\bar{v}$, increases in absolute value with the Bingham number. We find that the von Kármán pumping rate, $-\bar{w}_\infty$, is decreased for increasing values of the yield stress, this being a direct consequence of the reduction in the peak of the radial velocity profile. Since, in this case, the velocity functions decay to the far-field exponentially one finds that

$$\bar{\mu} \to 1 + \frac{2B_n e^{-\bar{w}_\infty \eta}}{\bar{w}_\infty \sqrt{A^2 + B^2}} + \cdots \quad \text{as} \quad \eta \to \infty,$$

whereA and B are equivalent constants. Therefore for $B_n>0$ the viscosity functions grow exponentially in the far-field, given that $\bar{w}_\infty \leqslant 0$, as observed in Fig. 2. This unphysical result owes from the inability of the Bingham model to describe apparent viscosity at vanishing shear-rates, as noted by Zhu et al. [13].

Comparative Results

A number of the previous comments regarding Bingham plastic fluids have also been noted by AS who considered the full system of non-linear governing equations by numerically integrating a three parameter system. Because of the boundary-layer formulation of this problem the governing Eq. (13a), (13b), (13c), (13d), (13e),(13f) and (13g) introduced here are reduced to a one parameter system, dependent only on B_n. In order to validate our results we compare our solutions with those of AS. Having retained all of the viscous terms, AS have a three parameter system dependent on[1]

$$B_y = \frac{\tau_y^*}{\mu_p^* \Omega^*},$$

Re and *r*, thus AS must specify values for all of the parameters before a numerical solution can be obtained.

However, recalling (9a) we see that B_n can be expressed simply as a function of r, *Re* and B_y. Hence we are able to construct comparative solutions given the data used by AS, that being r=1 and *Re*=2950. AS have obtained solutions for B_y=0,10,20,30,50(for r=1 and *Re*=2950) in a stationary frame of reference. In order to produce comparative results we transform our system from a rotating reference frame to a stationary one and solve the resulting equations.

Here we have reproduced the numerical solutions of AS using their three parameter numerical scheme. The results of these calculations are presented in Fig. 5. The solid line curves are the solutions obtained from the work of AS whilst the $*$ markers represent the solutions owing from the boundary-layer formulation of the problem. We note here that both sets of solutions decay to the far-field exponentially, as was outlined previously for the one parameter system.

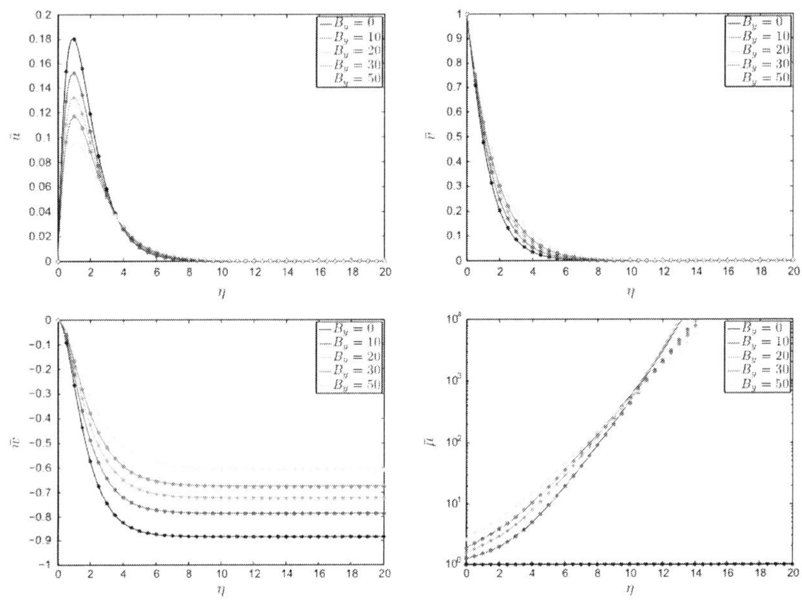

Figure 5: Plots of $\bar{u}, \bar{v}, \bar{w}$ and $\bar{\mu}$ versus η for Bingham plastic fluids with B_y=0,10,20,30,50. The solid line curves are reproductions of Ahmadpour and Sadeghy's solutions, the $*$ markers are the corresponding boundary-layer results.

It is clear from Fig. 5 that our results are in excellent agreement with those of AS. We find that there is in fact negligible discrepancy between the two sets of solutions. This serves to confirm the boundary-layer approximation made previously. We see that the elimination of the higher order viscous terms from the problem has little to no effect on the steady mean flow solutions.

In Fig. 3 and Fig. 4 basic flow solutions are plotted for shear-thinning and shear-thickening Carreau fluids, respectively. For both cases we find that the solutions have fully converged within the confines of the boundary-layer region $0 \leqslant \eta \leq 20$, which is in stark contrast to the results for shear-thinning power-law fluids presented in Fig. 1. As the power-law index increases from n=0.25 we observe that the peak in the radial cross-flow profile is shifted along the η-axis, indicating that the boundary-layer thickness increases with n. As such, the von Kármán pumping rate also increases with increasing $n$and does so in a non-linear fashion, see Fig. 6. We note the quantitative difference between power-law and Carreau shear-thinning solutions.

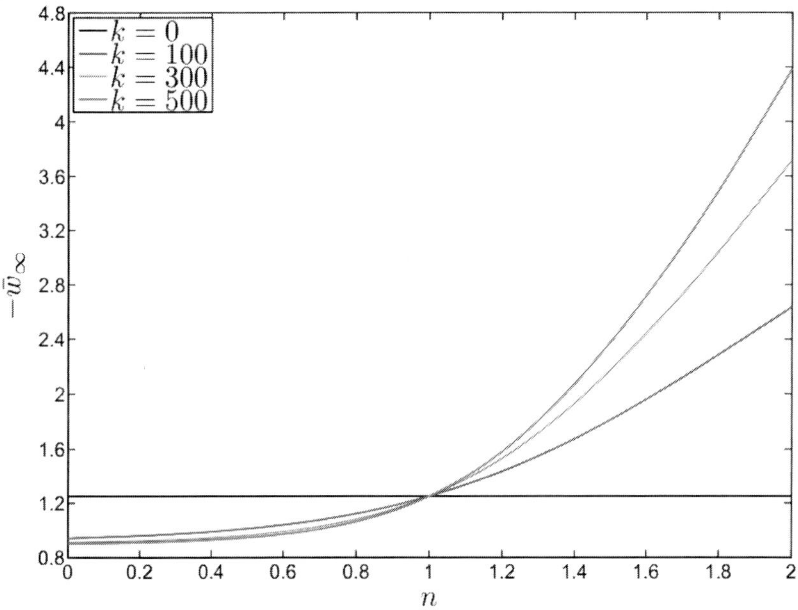

Figure 6: Variation of $-\bar{w}_\infty$ versus n for Carreau fluids with k=0,100,300,500.

Clearly as $\eta \to \infty$ we have that $\bar{\mu} \to 1 + c_0$ for all values of n and k. Thus, the Carreau model predicts a finite value of viscosity for both shear-thinning and shear-thickening fluids, where, in this case, the limit $1+c_0$ is approached from below and above, respectively. Furthermore, unlike for power-law and Bingham plastic fluids, we are able to accurately model the variation of the viscosity throughout the entirety of the boundary-layer. As noted by Dabrowski [12] this is due to the unrestricted ability of the Carreau model to describe non-Newtonian fluid behaviour for both $\dot{\gamma}^* \ll 1$ and $\dot{\gamma}^* \gg 1$.

Fig. 7 shows the effect of the 'relaxation' parameter, k, on the axial velocity and viscosity profiles. For shear-thinning (n=0.5) fluids, increasing the value of *k* has a rather minimal effect on the steady mean flow, this is observed in Table 2 as well as Fig. 7. Conversely, we see that increasing the value of k has rather a significant effect on both the velocity and viscosity profiles for shear-thickening (n=1.5) fluids. In this case both the von Kármán pumping rate and the boundary-layer thickness increases with *k*. Similar qualitative observations are noted by Dàbrowski [12] in his Falkner–Skan boundary-layer study.

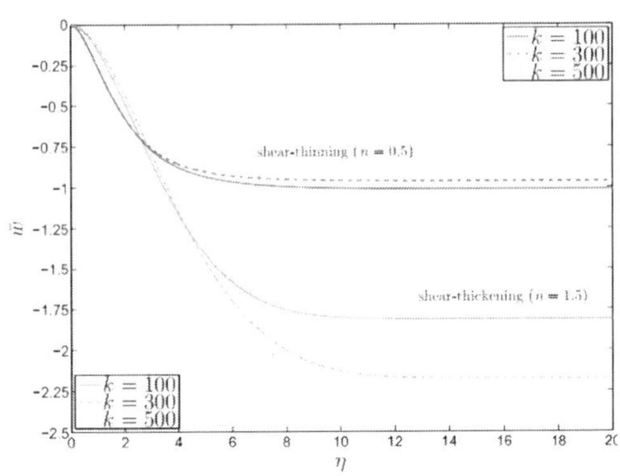

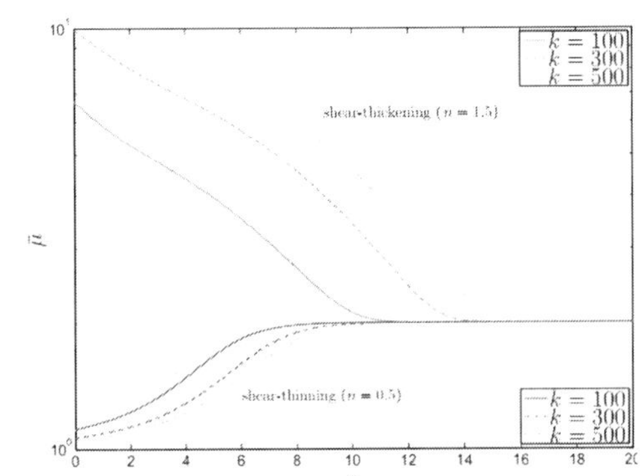

Figure 7: Variation of $\bar{w}$ and $\bar{\mu}$ versus η for shear-thinning and shear-thickening Carreau fluids withk=100,300,500.

CONCLUSION

Solutions for the boundary-layer flow on a rotating disk have been investigated for a number of generalised Newtonian fluid models. By firstly applying a large Reynolds number boundary-layer approximation, followed by the introduction of a von Kármán-like similarity solution, the governing partial differential equations are reduced to a set of five first-order ordinary differential equations. These are solved using a fourth order Runge–Kutta quadrature routine twinned with a Newton iteration scheme. The results for shear-thinning power-law fluids are consistent with those of Denier and Hewitt [8], whilst for Bingham plastic fluids we find that implementation of the boundary-layer approximation has a negligible effect on the solutions when compared to the full-field results of AS.

Interestingly, we note that the base flow similarity solution is still applicable when considering fluids with a more complex constitutive viscosity relationship, such as the Carreau model. In this case the structure of the flow is governed by the viscosity ratio $_{c0}$, the 'relaxation' parameter k and the power-law index n. As noted in Section 4 the Carreau model does not suffer from any of the non-physical flaws one encounters when employing either the power-law or Bingham plastic viscosity models, and as such is applicable at all shear-rates. Therefore, we suggest that the results presented within Section 3 provide a much better representation of the nature of the boundary-layer flow for both shear-thickening and shear-thinning non-Newtonian fluids. Dabrowski [12]reached a very similar conclusion in the context of flat plate boundary-layer flows. Griffiths et al. [14] and [15] have indicated that shear-thinning fluids may have a stabilising effect on the rotating disk boundary-layer flow. A natural extension of this work would be to see if such results are reproduced when considering the Carreau fluid model as opposed to the simpler power-law model. Has this predicted stabilising nature been simply a consequence of the failings of the power-law model at low shear-rates?

The stability of the flow for Bingham plastic fluids could also be investigated, at least asymptotically, as in the study of Griffiths et al. [14]. However, any attempt to implement the numerical integration scheme outlined in Griffiths et al. [15] would prove futile as this scheme integrates from the outer-edge of the boundary-layer down towards the surface of the disk. Here we have shown that the viscosity profiles grow exponentially in the far-field and thus the aforementioned numerical scheme would fail to initiate. In this case at least, a differing methodology would have to be adopted in order to construct curves of neutral stability. An indeed initial investigation into both of these problems has begun; reports on these studies are to be expected in due course.

It must be stated that in the absence of experimental validation the results presented in this study must be considered as theoretical predictions only. To the best of the author's knowledge no such experiments have yet taken place, suggesting that this is an area that requires future investigation.

The one-dimensional theoretical coating investigation of Jenekhe and Schuldt [16] may be of some interest here. In this study results are presented for film thickness profiles on a rotating disk for Newtonian, power-law and Carreau fluids. The authors conclude that the breakdown of the power-law model at vanishing shear-rates has a significant effect on the predicted film thickness. As with this study the results for power-law and Carreau fluids are in stark contrast to each other. This further suggests that the applicability of the power-law model must be questioned in the context of rotating flows of this nature.

ACKNOWLEDGMENTS

The author would like to thank Professor J.P. Denier for suggesting the Carreau fluid problem and gratefully acknowledges the support of the Engineering and Physical Sciences Research Council UK for his PhD studies.

Appendix A: Generalised Viscosity Terms

The additional terms omitted from (8h) are given here as

$$\mathcal{L}_{\mu} = \frac{1}{Re^{2/(q+1)}}\left\{2\left[\left(\frac{\partial u}{\partial r}\right)^2 + \left(\frac{u}{r}\right)^2 + \left(\frac{\partial w}{\partial z}\right)^2\right] + \left[r\frac{\partial}{\partial r}\left(\frac{v}{r}\right)\right]^2 + 2\frac{\partial u}{\partial z}\frac{\partial w}{\partial r}\right\} + \frac{1}{Re^{4/(q+1)}}\left(\frac{\partial w}{\partial r}\right)^2. \tag{A.1}$$

Appendix B: Pressure Distribution

The $\bar{v}$ terms in (13d) are given by

$$\text{Power-law model} - \quad \bar{v} = 2\bar{\mu}(q-1)/(q+1). \tag{B.1a}$$

$$\text{Bingham model} - \quad \bar{v} = 1 - \bar{\mu}, \tag{B.1b}$$

$$\text{Carreau model} - \quad \bar{v} = 2\bar{\mu}'/[\ln(\bar{u}'^2 + \bar{v}'^2)]'. \tag{B.1c}$$

As noted in Section 2 the pressure correction term, $\bar{p}$, is realisable once the velocity functions have been calculated. The profiles in Fig. B.8 have been determined using a standard trapezoidal numerical integration scheme where the unit step-size has been reduced until sufficiently converged results were obtained. In the Newtonian limit this scheme was validated against the analytic solution, $\bar{p} - \bar{p}_0 = -2\bar{u} - \bar{w}^2/2$, here we stipulate that $\bar{p}(\eta = 0) = 0 \Rightarrow \bar{p}_0 = 0$.

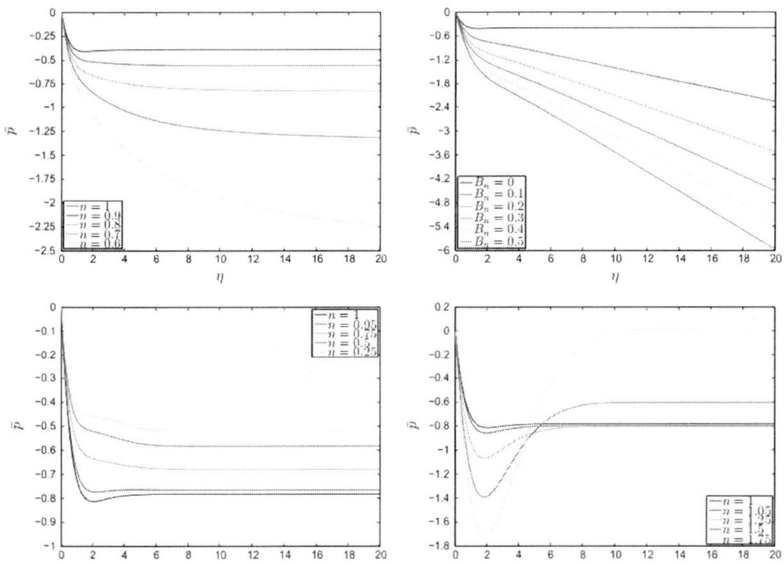

Figure B.8: Plots of $\bar{p}$ versus η for power-law, Bingham plastic and Carreau fluids.

Results for power-law and Carreau fluids are presented in Table B.3 and Fig. B.8. Due to the exponential growth of the viscosity function for Bingham plastic fluids we find that $\bar{p}$ is unbounded within the confines of the boundary-layer for B_n>0, as observed in Fig. B.8. Consequently we are unable to provide data for the limiting value of $\bar{p}$ in this case. As before this result is attributed to limitations of the Bingham model as $\dot{\gamma}^* \to 0$.

Table B.3: Numerical values of the pressure function at the outer-edge of the boundary-layer for Newtonian, power-law and Carreau fluids. The value of η_∞ employed for each power-law calculation is included. Solutions for n=1 are also included for Carreau fluids, these differ from the Newtonian results as in this case the viscosity function is effectively set to $\bar{\mu} = 1 + c_0 = 2$, rather than unity. Where not otherwise stated η_∞=20

Newtonian fluids	$-\bar{p}_\infty$	
$n - 1 = B_n = c_0 = 0$	0.3911	
Power-law fluids: $n =$	η_∞	$-\bar{p}(\eta_\infty)$
0.9	50	0.5565
0.8	100	0.8317
0.7	200	1.3698
0.6	400	2.7667
Carreau fluids: $k = 100, n =$	$-\bar{p}_\infty$	
0.25	0.5196	
0.5	0.5829	
0.75	0.6788	
0.95	0.7643	
1	0.7823	
1.05	0.7971	
1.25	0.8045	
1.5	0.6038	
1.75	−0.0120	

REFERENCES

1. T. von Kármán, Über laminare und turbulente Reibung, Z. Angew. Math. Mech. 1 (1921) 233–252.
2. G.K. Batchelor, Note on the class of solutions of the Navier–Stokes equations representing steady non-rotationally symmetric flow, Quart. J. Mech. Appl. Math. 4 (1951) 29–41.
3. U.T. Bödewadt, Die Drehströmung über festem Grund, Z. Angew. Math. Mech. 20 (1940) 241–252.

4. P.J. Zandbergen, D. Dijkstra, Von Kármán swirling flows, Annu. Rev. Fluid Mech. 19 (1987) 465–491.
5. P. Mitschka, Nicht-Newtonsche Flüssigkeiten II. Drehstrom¨ ungen Ostwald-de Waelescher Nicht-Newtonsche Flüssigkeiten, Coll. Czech. Chem. Commun. 29 (1964) 2892–2905.
6. P. Mitschka, J. Ulbrecht, Nicht-Newtonsche Flüssigkeiten IV. Strömung Nichtnewtonsche Flüssigkeiten Ostwald-de Waelescher Typs in der Umgebung Rotierender Drehkegel und Schieben, Coll. Czech. Chem. Commun. 30 (1965) 2511–2526.
7. H.I. Andersson, E. de Korte, R. Meland, Flow of a power-law fluid over a rotating disk revisited, Fluid Dyn. Res. 28 (2001) 75–88.
8. J.P. Denier, R.E. Hewitt, Asymptotic matching constraints for a boundary-layer flow of a power-law fluid, J. Fluid Mech. 518 (2004) 261–279.
9. A. Ahmadpour, K. Sadeghy, Swirling flow of Bingham fluids above a rotating disk: an exact solution, J. Non-Newtonian Fluid Mech. 197 (2013) 41–47.
10. S. Matsumoto, Y. Takashima, T. Kamiya, A. Kayano, Y. Ohta, Film thickness of a Bingham liquid film on a rotating disk, Ind. Eng. Chem. Fundam. 21 (1982) 198–202.
11. W.G. Cochran, The flow due to a rotating disk, Proc. Camb. Philos. Soc. 30 (1934) 365–375.
12. P.P. Dabrowski, Boundary-layer Flows in Non-Newtonian Fluids, Ph.D. thesis, School of Mathematical Sciences, The University of Adelaide, 2004.
13. H. Zhu, Y.D. Kim, D. De Kee, Non-Newtonian fluids with a yield stress, J. NonNewtonian Fluid Mech. 129 (2005) 177–181.
14. P.T. Griffiths, S.O. Stephen, A.P. Bassom, S.J. Garrett, Stability of the boundary layer on a rotating disk for power law fluids, J. Non-Newtonian Fluid Mech. 207 (2014) 1–6.
15. P.T. Griffiths, S.J. Garrett, S.O. Stephen, The neutral curve for stationary disturbances in rotating disk flow for power-law fluids, J. Non-Newtonian Fluid Mech. 213 (2014) 73–81.
16. S.A. Jenekhe, S.B. Schuldt, Coating flow of non-Newtonian fluids on a flat rotating disk, Ind. Eng. Chem. Fundam. 23 (1984) 432–436.

CITATION

P.T. Griffiths, Flow of a Generalised Newtonian Fluid due to a Rotating Disk, doi:10.1016/j.jnnfm.2015.03.008.

CHAPTER 7

Physical Modeling of Gas Pollutant Motion in the Atmosphere

OndrejZavila

[1]Vysoka SkolaBanska, Technical University of Ostrava, Faculty of Safety Engineering, Ostrava, Vyskovice, Czech Republic

INTRODUCTION

Air pollution is becoming an increasingly serious global issue. Factories produce large amounts of pollutants that damage the environment and harm human health. From this point of view, problems of motion and dispersion of pollutants in the atmosphere relate not only to environmental studies but also to other disciplines, such as safety engineering.

An understanding of the physical principles of pollutants' motion and dispersion is important in order to determine the impact of air pollution on the environment and humans. This study only deals with these physical principles of pollutants' motion and dispersion. Possible chemical reactions in the atmosphere are not covered.

For the purpose of the study, a simple model of a typical real situation was defined. Physical parameters of the model were gradually modified to achieve visible changes in results so that general principles could be defined. The above-mentioned demonstration model represents a chimney situated in a simple flat terrain. Gas pollutant is discharged from the chimney and carried by flowing air. Gas pollutant plume is detected and visualized with a numerical model as iso-surfaces or contours of pollutant concentrations in two-dimensional cut planes of three-dimensional geometry.

The dependence of the pollutant plume shape, size and inclination on modification of three physical parameters was investigated. The selected parameters included pollutant density, air flow velocity and model scale.

The results are presented in the form of text, commented figures and tables. ANSYS Fluent 13.0 CFD (Computational Fluid Dynamics) code was used to demonstrate and visualize all problem variants (see [1],[2]). The numerical model of the pollutant plume motion created in this software had been verified by an experiment conducted in the low-speed wind tunnel in the Aerodynamic Laboratory of the Academy of Sciences of the Czech Republic in NovyKnin (see [3],[4],[5]). One of the aims of the study is also to demonstrate that physical modeling of pollutant plume motion and dispersion with severely downscaled models has its limitations that should be known and considered to avoid obtaining false results.

PHYSICAL MODELING IN WIND TUNNELS

Wind tunnels are facilities where specific air flow regime can be set up with a certain level of precision as required by investigators. They are primarily used to study aerodynamic characteristics of bodies. The history of wind tunnels dates back to 1751 and is connected with the name of Francis H. Wenham, a Council Member of the Aeronautical Society of Great Britain. In the beginning, wind tunnels were used mainly in aviation development. Nowadays, they are used for many applications in, for instance, automotive industry, building industry, environmental and safety engineering.
Wind tunnels can be classified, for example, by the velocity of air flow (see [6]) as:

a. *Low-speed wind tunnels* – air flow velocity in the measuring section of the tunnel is low enough to consider the air as an incompressible medium;
b. *High-speed wind tunnels* – air flow velocity in the measuring section of the tunnel is high enough to consider the air as a compressible medium;

c. *Subsonic wind tunnels* - air flow velocity in the measuring section of the tunnel is high enough (subsonic) to consider the air as a compressible medium;
d. *Supersonic wind tunnels* - air flow velocity in the measuring section of the tunnel is high enough (supersonic) to consider the air as a compressible medium.

Except for applications using only clean air, there are many applications in which various pollutants are applied. Pollutants are introduced to visualize the air flow field or to simulate the motion and dispersion of pollution in the real atmosphere. These applications can be found, for example, in environmental and safety engineering. Results of measurements using downscaled models in wind tunnels can be used as base data for local emergency planning studies.

Criteria of Physical Similarity

Two phenomena can be considered to be similar (despite different geometrical scales) if three types of similarity match: geometric, kinematic and dynamic. Criteria of geometric similarity require that the ratios of main corresponding dimensions on the model and the original pattern be constant. Also, main corresponding angles on the original pattern and the model must be of the same value. Criteria of kinematic similarity require that the ratios of velocities at corresponding points be the same for both the original pattern and the model. Criteria of dynamic similarity require that the ratios of the main forces at corresponding points be the same for both the original pattern and the model.

Forces can be divided into two groups: areal forces and volume (weight) forces. Areal forces include friction forces, compression forces, and capillary (surface) forces. Volume (weight) forces include inertial forces, gravity forces and impulse forces (resulting from the change in momentum). According to the type of phenomena, these forces can be put into mathematical relation and criteria (numbers) of similarity can be established. In fluid mechanics, the Reynolds number, Euler number, Newton number, Froude number, Weber number, and Mach number are the most widely known criteria. Each of them expresses ratio between two different forces. In practice, it is not possible to achieve correspondence between the original pattern and the model in all criteria. Therefore, it is always up to the investigators who must use their knowledge and experience to choose the right and most important criterion (or criteria) for the investigated phenomenon. As a result, investigators usually work with one or two dominant criteria of similarity [7],[8],[9].

Applying Froude Number

The Froude number expresses the ratio between gravity forces and inertial forces. Gravity forces cause vertical movements of the plume (climbing or descending) and inertial forces cause horizontal movements of the plume. The Froude number can be therefore considered as a criterion of dynamic similarity, which should be of the same value for both the scaled model and the real pattern (see [7],[8],[9]).

The Froude number can be defined as

$$Fr \approx \frac{F_{I\ air}}{F_{G-pollutant}} = \frac{\rho_{air} \cdot S \cdot v_{air}^2}{\rho_{pollutant} \cdot g \cdot V} = \frac{\rho_{air} \cdot l_2 \cdot l_3 \cdot v_{air}^2}{\rho_{pollutant} \cdot g \cdot l_1 \cdot l_2 \cdot l_3} = \frac{\rho_{air} \cdot v_{air}^2}{\rho_{pollutant} \cdot g \cdot l_1} \qquad (1)$$

where F_{I-air} is the inertial force due to the air acting on pollutant element [N], F_G−pollutant is the gravity force acting on pollutant element [N], ρ air is the air density [kg/m^3], $\rho_{pollutant}$ is the pollutant density [kg/m^3], S is the surface of pollutant element acted on by the flowing air [m^2], v_{air} is the air flow velocity [m/s], g is the gravity acceleration constant [m/s^2], V is the volume of pollutant element released from the pollutant source per 1 second [m^3], l_1 is the 1st characteristic dimension of the pollutant source (length of pollutant cubic element) [m], l_2 is the 2nd characteristic dimension of the pollutant source (width of pollutant cubic element) [m] and l_3 is the 3rd characteristic dimension of the pollutant source (height of pollutant cubic element) [m]. l_3 can be replaced by $v_{pollutant}$ that represents the velocity of the pollutant released from the source in vertical direction [m/s]. Fr is a dimensionless constant [-] whose value determines whether the inertial force or the gravity force will dominate in the specific pollutant plume motion scenario.

All important characteristics are illustrated in Figure 1. Pollutant element was simplified into a rectangular cuboid with dimensions of l_1, l_2 and l_3 to make practical calculations easier. Of course, the spout of a real chimney can be of a different shape, most commonly circular or elliptical. In this case it is advisable to calculate the surface of the spout and transform the shape into a square or a rectangle with dimensions of l_1 and l_2. The value of l_3 remains the same (despite the shape of the spout) and is replaced by the velocity with which the pollutant leaves the source in vertical direction v pollutant.

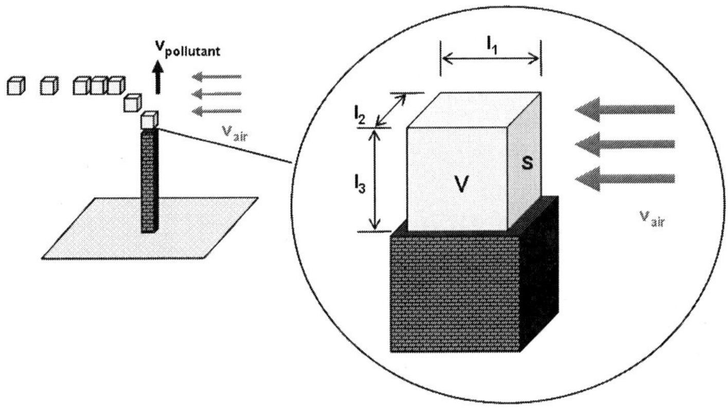

Figure 1: Air flow action on gas pollutant element leaking from the nozzle (chimney).

If Fr< 1, gravity forces are assumed greater than inertial forces. Hence, vertical motions (climbing or descending) of the gas pollutant plume can be expected due to different densities of the pollutant and the air. Plumes of light gas pollutants will tend to climb, whereas plumes of heavy gas pollutants will tend to descend.

If Fr = 1, gravity forces are assumed equal to inertial forces. Hence, gas pollutant plumes are carried by flowing air along with manifesting partial vertical motions.

If Fr> 1, inertial forces are assumed greater than gravity forces. Hence, vertical motions of the gas pollutant plume are limited or nonexistent. The gas pollutant plume is carried by strong flowing air, regardless of the pollutant—air density difference or weight of the pollutant.

This third scenario causes common difficulties when planning gas pollutant plume motion and dispersion experiments with downscaled models in low-speed wind tunnels. At small dimensions of measuring sections of common wind tunnels and, thus, low scales of models, the air flow may be too great to allow vertical motions of gas pollutant plumes. Proper conditions often cannot be assured in such cases.

Influence of Air Velocity

According to Equation (1) in Section 1.2, the air flow velocity v_{air} influences the inertial force F_{I-air} that causes gas pollutant horizontal motion. If all the other physical characteristics are constant, the following principles can be formulated: The greater the air flow velocity v_{air} is, the greater the inertial force F_{I-air} is. The greater the inertial force is, the more limited the pollutant plume vertical motions (climbing or descending) are.

The air flow velocity $v_{air,Fr=1}$ for Fr = 1 (i.e., inertial and gravity forces are equal) can be deduced from Equation (1) mentioned in Section 1.2:

$$1 = \frac{F_{I-air}}{F_{G-pollutant}} = \frac{\rho_{air} \cdot S \cdot v_{air,Fr=1}^2}{\rho_{pollutant} \cdot g \cdot V} = \frac{\rho_{air} \cdot l_2 \cdot l_3 \cdot v_{air,Fr=1}^2}{\rho_{pollutant} \cdot g \cdot l_1 \cdot l_2 \cdot l_3} = \frac{\rho_{air} \cdot v_{air,Fr=1}^2}{\rho_{pollutant} \cdot g \cdot l_1} \tag{2}$$

Thus, the air flow velocity $v_{air,Fr=1}$ is

$$v_{air,Fr=1} = \sqrt{\frac{\rho_{pollutant} \cdot g \cdot l_1}{\rho_{air}}} = \sqrt{\frac{\rho_{pollutant} \cdot g \cdot l_1 \cdot l_2 \cdot l_3}{\rho_{air} \cdot l_2 \cdot l_3}} = \sqrt{\frac{\rho_{pollutant} \cdot g \cdot V}{\rho_{air} \cdot S}} \tag{3}$$

However, one must realize that the change in air flow velocity influences also air flow field turbulent characteristics. For example, turbulent intensity is influenced when the air flows around solid objects or in a complex terrain. Investigators must consider whether these changes have a serious impact on accuracy of the experiment or mathematical model. This is very important for modeling gas pollutant motion and dispersion in a complex geometry (complex terrain) where the real model of turbulent flow field is the key element of the simulation. If the air turbulent flow field is seriously influenced by the change in the air flow velocity, the results of the analysis can be misleading. This approach is therefore not suitable for such cases and a different parameter of the model must be changed (see Sections 1.4 and 1.5).

Influence of Pollutant Density

According to Equation (1) in Section 1.2, the pollutant density $\rho_{pollutant}$ influences the gravity force $F_{G-pollutant}$ that causes gas pollutant vertical motions (climbing or descending). If all the other physical characteristics are constant, the following principles can be formulated: If the pollutant density $\rho_{pollutant}$ is greater than the air density ρ_{air}, the pollutant tends to descend (i.e., the gas pollutant plume descends). If the pollutant density $\rho_{pollutant}$ is lower than the air density ρ_{air}, the pollutant tends to climb (i.e., the gas pollutant plume climbs). The greater the gravity forces $F_{G-pollutant}$ are, the more significant the gas pollutant plume vertical movements (climbing or descending) are.

The pollutant density $\rho_{pollutant,Fr=1}$ for Fr= 1 (i.e., inertial and gravity forces are equal) can be deduced from Equation (1) in Section 1.2:

$$1 = \frac{F_{I-air}}{F_{G-pollutant}} = \frac{\rho_{air} \cdot S \cdot v_{air}^2}{\rho_{pollutant,Fr=1} \cdot g \cdot V} = \frac{\rho_{air} \cdot l_2 \cdot l_3 \cdot v_{air}^2}{\rho_{pollutant,Fr=1} \cdot g \cdot l_1 \cdot l_2 \cdot l_3} = \frac{\rho_{air} \cdot v_{air}^2}{\rho_{pollutant,Fr=1} \cdot g \cdot l_1} \quad (4)$$

Thus, the pollutant density $\rho_{pollutant,Fr=1}$ is

$$\rho_{pollutant,Fr=1} = \frac{\rho_{air} \cdot v_{air}^2}{g \cdot l_1} = \frac{\rho_{air} \cdot v_{air}^2 \cdot l_2 \cdot l_3}{g \cdot l_1 \cdot l_2 \cdot l_3} = \frac{\rho_{air} \cdot v_{air}^2 \cdot S}{g \cdot V} \quad (5)$$

A change in pollutant density in order to achieve the optimum ratio between inertial and gravity forces would be often the ideal solution. However, there is a problem. The densities of pollutants range within a narrow interval—approximately of one of order of magnitude—which is usually is not enough to compensate the

Froude number differences resulting from, e.g., a substantial change of the model scale.

A typical example can be the physical modeling of gas pollutant plumes in low-speed wind tunnels where the scale of the model is around 1:1000. In such a case, there is a need to change the pollutant density 100, or even 1000 times, which is impossible. This is why change in pollutant density can be used to achieve only a small change in the Froude number. These small changes, however, may not be sufficient for a successful execution of the experiment or numerical modeling.

The change of the flowing gas density could be an alternative solution. For example, air could be replaced by a different gas with a different value of density. However, this change influences turbulent flow field characteristics, which may be undesirable (see Section 1.3).

Influence of Model Scale

According to Equation (1) in Section 1.2, the model scale can be expressed by using the value l_1 that represents the 1st characteristic dimension of the pollutant source (i.e., length of pollutant cubic element) (see Figure 1). The model scale influences both inertial and gravity forces and changes their ratio. If all the other physical characteristics are constant, the following principles can be formulated: The greater the model scale is, the greater the influence of gravity forces is. The smaller the model scale is, the greater the influence of inertial forces is. Gravity forces are, e.g., greater in a model scaled at 1:4 than in that scaled at 1:1000 (see Section 2.6).

The value of the 1st characteristic dimension of the pollutant source (length of pollutant cubic element) $l_{1,Fr=1}$ for Fr = 1 (inertial and gravity forces are equal) can be deduced from Equation (1) in Section 1.2:

$$1 = \frac{F_{I-air}}{F_{G-pollu\tan t}} = \frac{\rho_{air} \cdot S_{Fr=1} \cdot v_{air}^2}{\rho_{pollu\tan t} \cdot g \cdot V_{Fr=1}} = \frac{\rho_{air} \cdot l_{2,Fr=1} \cdot l_{3,Fr=1} \cdot v_{air}^2}{\rho_{pollu\tan t} \cdot g \cdot l_{1,Fr=1} \cdot l_{2,Fr=1} \cdot l_{3,Fr=1}} \quad (6)$$

or

$$1 = \frac{F_{I-air}}{F_{G-pollu\tan t}} = \frac{\rho_{air} \cdot v_{air}^{2}}{\rho_{pollu\tan t, Fr=1} \cdot g \cdot l_{1,Fr=1}} \tag{7}$$

Thus,

$$l_{1,Fr=1} \tag{8}$$

is

$$l_{1,Fr=1} = \frac{\rho_{air} \cdot v_{air}^{2}}{\rho_{pollu\tan t} \cdot g} = \frac{\rho_{air} \cdot v_{air}^{2} \cdot l_{2,Fr=1} \cdot l_{3,Fr=1}}{\rho_{pollu\tan t} \cdot g \cdot l_{2,Fr=1} \cdot l_{3,Fr=1}} = \frac{\rho_{air} \cdot v_{air}^{2} \cdot S_{Fr=1}}{\rho_{pollu\tan t} \cdot g \cdot S_{Fr=1}} \tag{9}$$

T model scale $M_{Fr=1}$[-] for Fr= 1 (inertial and gravity forces are equal) is given by

$$M_{Fr=1} = \frac{1}{X} \tag{10}$$

where

$$X = \frac{l_1}{l_{1,Fr=1}} \tag{11}$$

The value l1 represents the 1st characteristic dimension of the pollutant source (length of pollutant cubic element) in the original model. The value $l_{1,Fr=1}$ is the 1st characteristic dimension of the pollutant source in the model where Fr= 1 (i.e., inertial and gravity forces are equal).

If the model scale is changed, the change of the inertial force is given by

$$F_{I1} = X^2 \cdot F_{I2} \tag{12}$$

and the change of the gravity force is given by

$$F_{G1} = X^3 \cdot F_{G2} \tag{13}$$

The value X is a model scale factor [-]. If X> 1, the model is smaller than its original pattern (i.e., the model is downscaled). If X< 1, the model is larger than its original pattern (i.e., the model is enlarged).Equations (11) and (12) are deduced from Equation (1) for the Froude number. The situation is illustrated in Figure 2.

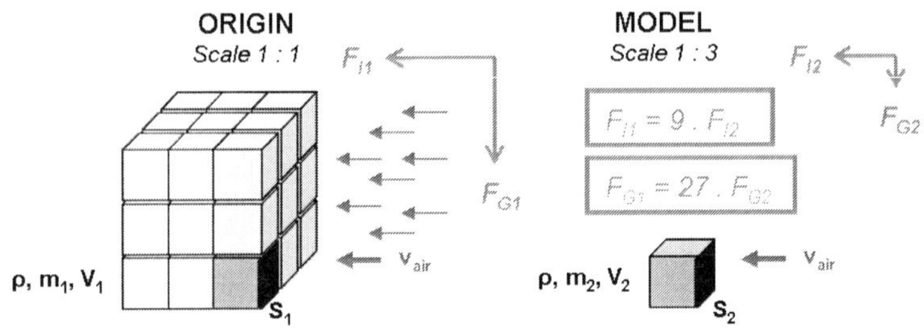

Figure 2: Inertial and gravity forces acting on gas pollutant element in two different model scales.

A change in model scale always causes a change in the ratio of inertial and gravity forces. Therefore, some problems cannot be reliably modeled at other than approximately original scales. The range of deviation from the original depends on the discretion of investigators. Investigators must decide whether the tolerance of the results is acceptable.

Modeling of gas pollutant plume motions with severely downscaled models is a typical example of this problem. Downscaled pollutant plume models will not correspond to the originally scaled patterns without modifying some key physical characteristics (air flow velocity, turbulent characteristics, etc.).

The problem can be solved by using a numerical mathematical model verified by a clearly defined experiment of the same type of physical phenomenon. Once the numerical model is verified, it can be used for numerical simulation of any problem of the same physical principles, whatever the model scale is. Sometimes it is impossible to do the same with a physical experiment.

It can be concluded that severely downscaled experiments are not suitable for modeling gas pollutant plume motions because of possible absence of vertical movements. It is more advisable to use physical experiment data only for verification of the numerical mathematical model (code, software).

CFD NUMERICAL MODELING OF GAS POLLUTANT MOTION

CFD (Computational Fluid Dynamics) is an abbreviation for a category of sophisticated mathematical codes for fluid mechanics computations. These codes offer a wide range of applications including modeling of turbulent fluid flows, heat transfers, and species motions with or without chemical reactions. It is an effective tool for prediction and reconstruction of many physical phenomena, especially in mechanical engineering, civil engineering, safety engineering, power engineering, environmental engineering, and in many other disciplines.

CFD Numerical Modeling Basics

CFD (Computational Fluid Dynamic) codes are based on the numerical solution of systems of partial differential equations that express the law of conservation of mass (continuity equation), the law of conservation of momentum (Navier-Stokes equations) and the law of conservation of energy (energy equation). This basic set of equations can be supplemented by additional equations that express heat transfer (heat transfer equations - convection, conduction or radiation), or species transport (species transport equations - gas, liquid or solid). The system of equations is then solved with an appropriate numerical method.

At the start of the modeling process, the geometrical shape of a two- or three-dimensional geometry is created that represents the object of modeling and its close surroundings. The size and shape of the geometry must enable the investigated phenomena to be captured.

In the second step, the geometry is covered with a grid. The grid divides the geometry into a finite number of two- or three-dimensional elements (cells) of a certain shape in which relevant physical quantities will be calculated. The quantities are calculated at the middle of the cells. Values in intermediate spaces are interpolated or extrapolated. The type and quality of grid influence the calculation time and the quality of results. A quality grid is smooth enough to capture the modeled phenomenon and its cells have as little shape deformation as possible compared to the ideal original shape.

In the third step, investigators choose the most suitable mathematical model and sub-models for the calculation of the problem. For gas pollutant plume motion and dispersion, the model of turbulence for the air flow field calculation is chosen first, than the species transport model is chosen for species transport modeling.

In the fourth step, boundary and initial conditions are defined. Boundary conditions are physical constants or variables that characterize the physical conditions at the boundaries of the geometry, i.e., at its inlet, outlet or walls. Boundary conditions do not change during the course of the calculation. They can be defined as a constant, a spatial-dependent function or a time-dependent function (polynomial, linear or periodic). Initial conditions are physical constants or variables that characterize the initial state of the system. Unlike boundary conditions, the initial conditions change in the course of the calculation after each iteration. If data for setting initial conditions are missing, at least approximate values should be entered. Initial conditions influence the calculation run-time, i.e., the time required to reach convergence.

In the fifth step, investigators set mathematical parameters of the calculation, e.g., solution schemes, under-relaxation parameters, and criteria of convergence. If required, as in case of calculations of time-dependent problems, special post-processing functions are set in this step, too.

Afterwards, the calculation is started and its progress monitored. Lastly, results are evaluated with available standard software tools or with preset post-processing functions (animations, time-dependent progress of physical variables, etc.) (see [1],[2]).

Basic Equations

The following equations are used for basic air flow field calculations:

Continuity Equation for Compressible Fluid Flow

The continuity equation expresses the law of conservation of mass. For unsteady (time-dependent) compressible fluid flows, it can be written as a differential equation in form of

$$\frac{\partial \rho}{\partial t} + \frac{\partial (\rho \cdot \bar{u}_j)}{\partial x_j} = 0 \tag{14}$$

where ρ is the fluid density [kg/m^3], t is time [s], $\bar{u}_j$ is the time-averaged j-coordinate of the fluid flow velocity [m/s], and x_j is a coordinate of the Cartesian coordinate system [-].

Equations for Compressible Fluid Flow

These equations—known as Navier-Stokes equations—express the law of conservation of momentum. For calculating turbulent flows, time-averaged values must be applied. The substitution of the time-averaged values into the Navier-Stokes equations gives the Reynolds equations. The equation of conservation of momentum for compressible fluids can be written in the form corresponding to that in differential form as

$$\frac{\partial (\rho \cdot \bar{u}_i)}{\partial t} + \frac{\partial (\rho \bar{u}_i \cdot \bar{u}_j)}{\partial x_j} = -\frac{\partial \bar{p}}{\partial x_i} + \frac{\partial}{\partial x_j} \cdot \tau_{ji} + \rho \cdot \delta_{i3} \cdot g + \rho \cdot f_c \cdot \varepsilon_{ij3} \cdot \bar{u}_j + \rho \cdot f_j \tag{15}$$

where ρ is the fluid density [kg/m^3], t is time [s], $\bar{u}_j$ is the time-averaged j -coordinate of the fluid flow velocity [m/s], x_j is a coordinate of the Cartesian coordinates system [-], $\bar{p}$ is the time-averaged value of pressure [Pa], τ_{ji} is the tensor of viscous stress [Pa], δ_{i3} is the Kronecker delta [-], ε_{ij3} is the unit tensor for centrifugal forces [-], f_j is the j -coordinate of force [N], and g is the gravity acceleration [m/s^2] (if gravity forces are included).
The meaning of terms in Equation (14):

1st term on the left	....	Unsteady term (acceleration)
2nd term on the left	....	Convective term
1st term on the right	....	Influence of pressure forces
2nd term on the right	....	Influence of viscous stress
3rd term on the right	....	Influence of gravity forces (buoyancy)
4th term on the right	....	Influence of Earth rotation (Coriolis force)
5th term on the right	....	Influence of volumetric forces

The equation of turbulent kinetic energy k and the equation of turbulent dissipation rate ε are used to express the turbulent flow field variables. The exact equation for k can be deduced from the Navier-Stokes equations and written as

$$\frac{\partial k}{\partial t} + \frac{\partial \overline{u_j} \cdot k}{\partial x_j} = -\frac{\partial}{\partial x_j} \cdot \left[\overline{u'_j \cdot \left(\frac{u'_l \cdot u'_l}{2} + \delta_{jl} \cdot \frac{p'}{\rho} \right)} \right] + \nu_t \cdot \frac{\partial^2 k}{\partial x_j^2} - \overline{u'_l \cdot u'_j} \cdot \frac{\partial \overline{u_l}}{\partial x_j} - \nu \cdot \overline{\frac{\partial u'_l}{\partial x_j} \cdot \frac{\partial u'_l}{\partial x_j}} \tag{16}$$

where k is the turbulent kinetic energy [m^2/s^2], t is time [s], $\overline{u}_j$ is the time-averaged j -coordinate of the fluid flow velocity [m/s], x_j is a coordinate of the Cartesian coordinate system [-], ρ is the fluid density [kg/m^3], p' is a component of the pressure fluctuation [Pa], and ν_t is the turbulent kinematic viscosity [Pa.s].

The meaning of terms in Equation (15):

1st term on the left	....	Turbulent energy rate
2nd term on the left	....	Convective transport of turbulent energy
1st term on the right	....	Turbulent diffusion due to pressure and velocity fluctuation
2nd term on the right	....	Molecular diffusion
3rd term on the right	....	Production of turbulent kinetic energy due to sliding friction
4th term on the right	....	Viscous dissipation

The turbulent kinetic energy k from Equation (15) is

$$k = \frac{1}{2} \cdot \left(\overline{u'^2_1} + \overline{u'^2_2} + \overline{u'^2_3} \right) = \frac{1}{2} \cdot \overline{u'^2_j} \tag{17}$$

where $\bar{u}_j$ represents time-averaged flow velocity components [m/s].

The exact equation for ε can be deduced from the Navier-Stokes equations and written as

$$\frac{\partial \varepsilon}{\partial t} + \frac{\partial \bar{u}_j \cdot \varepsilon}{\partial x_j} = \frac{\partial}{\partial x_j} \cdot \left(\frac{\nu_t}{\sigma_\varepsilon} \cdot \frac{\partial \varepsilon}{\partial x_j} \right) + C_{1\varepsilon} \cdot \nu_t \cdot \left(\frac{\partial \bar{u}_j}{\partial x_l} + \frac{\partial \bar{u}_l}{\partial x_j} \right) \cdot \frac{\partial \bar{u}_l}{\partial x_j} - C_{2\varepsilon} \cdot \frac{\varepsilon^2}{k} \tag{18}$$

where ε is the turbulent dissipation rate [m^2/s^3], t is time [s], $\bar{u}_j$ is the time-averaged j -coordinate of the fluid flow velocity [m/s], x_j is a coordinate of the Cartesian coordinate system [-], ν_t is the turbulent kinematic viscosity [m^2/s], σ_ε , $C_{1\varepsilon}$ and $C_{2\varepsilon}$ are empirical constants [-], and k is the turbulent kinetic energy [m^2/s^2].

The turbulent kinetic viscosity ν_t is

$$\nu_t = C_\nu \cdot \frac{k^2}{\varepsilon} \tag{19}$$

where C_ν is an empirical constant [-].

Energy equation

The energy equation expresses the law of conservation of energy. According to this law, change in the total fluid energy $\overline{E}$ in the volume V is determined by the change of the inner energy and kinetic energy, and by the flux of both energies through surface S [m^2] that surrounds volume V [m^3]. The final equation can be written as

$$\frac{\partial}{\partial t}\cdot\left[\rho\cdot\overline{E}\right]+\frac{\partial}{\partial x_j}\cdot\left[\rho\cdot\bar{u}_j\cdot\overline{E}\right]=\rho\cdot\bar{u}_j\cdot f_j-\frac{\partial\left(p\cdot\bar{u}_j\right)}{\partial x_j}+\frac{\partial\left(\tau_{ji}\cdot\bar{u}_j\right)}{\partial x_l}-\frac{\partial\bar{q}_j}{\partial x_j}\qquad(20)$$

where t is time [s], ρ is the fluid density [kg/m³], $\overline{E}$ is the time-averaged value of energy [J/kg], $\bar{u}_j$ is the time-averaged j-coordinate of the flow field velocity [m/s], x_j is a coordinate of the Cartesian coordinate system [-], p is the pressure [Pa], τ_{ji} is the tensor of viscous stress [Pa], and $\bar{q}_j$ is the time-averaged j-coordinate of heat flux [J/(m².s)].

The meaning of terms in Equation (19):

1st term on the left	....	Energy rate
2nd term on the left	....	Convective transport of energy
1st term on the right	....	Work of outer volumetric forces (gravity force)
2nd term on the right	....	Thermodynamic reversible energy flux into the volume V and its change caused by pressure forces
3rd term on the right	....	Irreversible growth of energy due to dissipation caused by viscosity
4th term on the right	....	Vector of heat flux according to the Fourier law

The following equation is used for the calculations of species transport:

Species Transport Equation

In the model, time-averaged values of the local species mass fraction $\overline{Y}_{i'}$ are calculated. These values are described by a balance equation similar to the energy equation (see Equation 19). The energy equation includes both convective and diffuse components of the transport. It can be written in conservative form as

$$\frac{\partial}{\partial t}\left(\rho\cdot\overline{Y}_{i'}\right)+\frac{\partial}{\partial x_j}\cdot\left(\rho\cdot\bar{u}_j\cdot\overline{Y}_{i'}\right)=-\frac{\partial}{\partial x_i}J_{j,i'}+R_{i'}+S_{i'} \tag{21}$$

where $\bar{u}_j$ is the time-averaged j -coordinate of the flow field velocity [m/s], R_i' is the production rate of species i' due to chemical reaction [kg/(m³.s)], and $S_{i'}$ is the growth production rate from distributed species [kg/(m³.s)]. The Equation (20) is valid for N−1 species, where N is the total number of components presented in the mathematical model [-]. Species distribution can be performed under various conditions, for which it can generally be divided into laminar and turbulent flow distribution. The value $J_{j,i'}$ represents the j -coordinate of i' -species diffuse flux [kg/(m³.s)]. To express the i' -species diffuse flux in the turbulent flow regime, the model employs

$$J_{i'}=-\left(\frac{\mu_t}{Sc_t}\right)\cdot\frac{\partial\overline{Y}_{i'}}{\partial x_j} \tag{22}$$

where μ_t is the turbulent dynamic viscosity [Pa.s], $\overline{Y}_{i'}$ is the time-averaged species i' mass fraction [-], and Sc_t is the Schmidt turbulent number [-] (set at the default value of 0.7) (see [1],[2]).

Example of Numerical Simulation – Gas Pollutant Motion in Wind Tunnel

ANSYS Fluent 13.0, one of the world's most sophisticated CFD codes, was chosen for the numerical simulation of the gas pollutant plume motion and dispersion. The gauging section of the low-speed wind tunnel with a small nozzle representing a chimney in a flat, simple terrain was the object of the numerical simulation. Gas pollutant enters the gauging section through the top of the nozzle (chimney) and is carried by flowing air (see Figure 3 to Figure 6).

Three-dimensional cubic geometry in five different model scales was designed for the purpose of this analysis (see Table 1).

The geometry grid at all modeled scales consisted of 569 490 three-dimensional cells. The problem was defined as steady (time-independent) with accuracy set at the value of 0.0001 (criterion of convergence).

RANS (Reynolds-averaged Navier-Stokes equations) approach was used for turbulent characteristics definition. RNG k−ε model of turbulence was used for the air flow field basic calculation. The Boussinesq hypothesis of swirl turbulent viscosity was applied for the turbulent viscosity calculation (see [1], [2]). Species transport model was used for the species motion calculation. Both models worked simultaneously. No additional gas pollutant dispersion model was applied. The operating pressure was set at 101 325 [Pa], the operating temperature was 300 [K], and the gravity acceleration -9.81 [m/s^2] in the geometry. Considering the pollutant source close surroundings, the ranges of the Reynolds number Re were 250–1250 [-] (model scale 1:1000, referential air flow velocity 1–5 [m/s], and nozzle spout diameter 0.0035 [m]) and 250000–1250000 [-] (model scale 1:1, referential air flow velocity 1–5 [m/s], and chimney spout diameter 3.5 [m]).

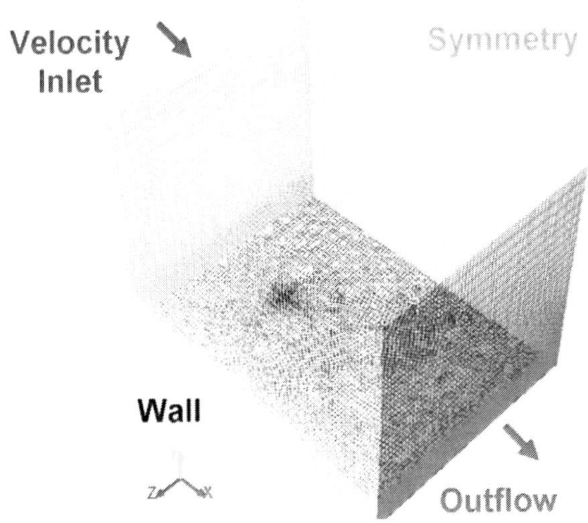

Figure 3: Geometry and grid (isometric view).

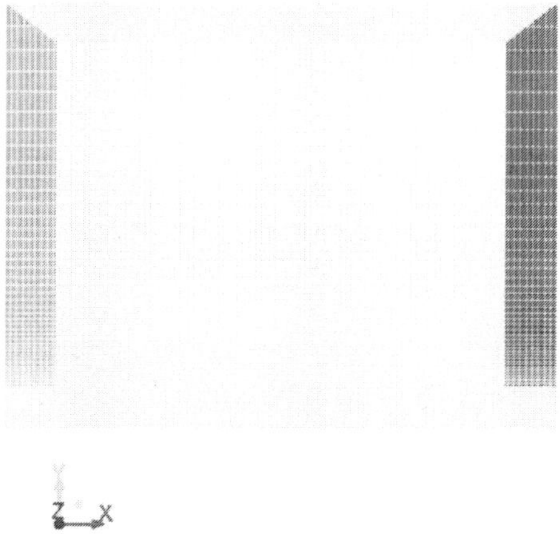

Figure 4: Geometry and grid (right side view).

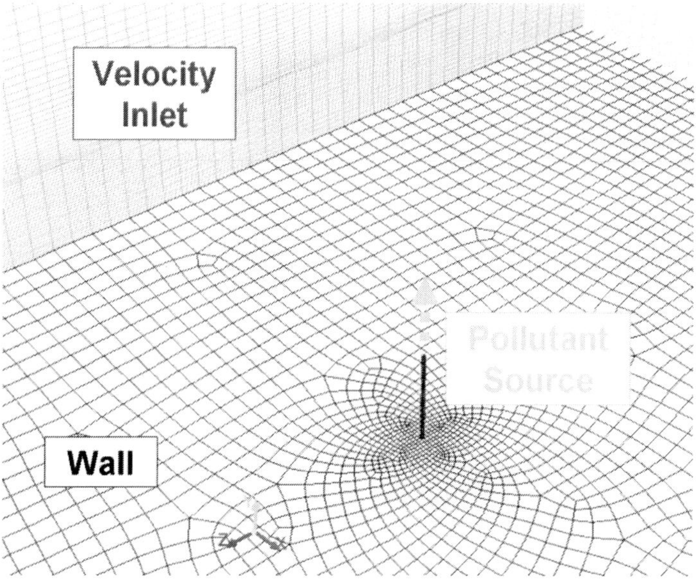

Figure 5: Pollutant source.

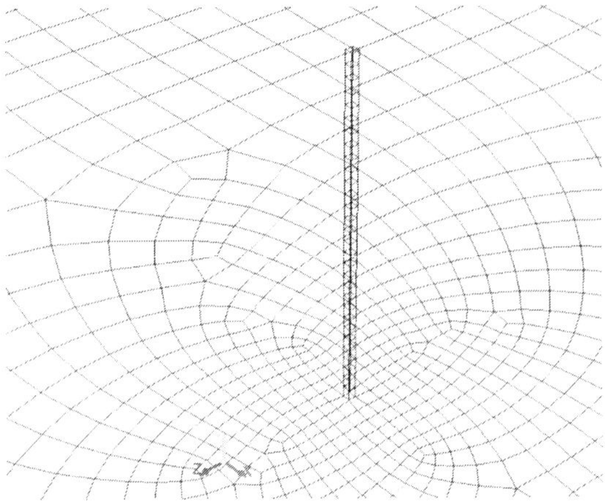

Figure 6: Pollutant source (detail).

Table 1: Geometry dimensions and pollutant source coordinates by model scale

Model Scale [-]	Geometry Dimensions [m]	Nozzle (Chimney) Dimensions [m]	Nozzle (Chimney) Spout Surface [m]	Coordinates of the Nozzle (Chimney) [m]	*Note*
1:01	X = 2000	X = 3.5	X = 3.5	X = 500	Real
	Y = 1500	Y = 200	Z = 3.5	Y = 0	Pattern
	Z = 1500	Z = 3.5		Z = 750	
01:04.0	X = 495.04	X = 0.866	X = 0.866	X = 123.76	-
	Y = 371.28	Y = 49.5	Z = 0.866	Y = 0	
	Z = 371.28	Z = 0.866		Z = 185.64	
01:35.5	X = 56.32	X = 0.098	X = 0.098	X = 14.08	-
	Y = 42.24	Y = 5.632	Z = 0.098	Y = 0	
	Z = 42.24	Z = 0.098		Z = 21.12	
02:41.8	X = 19.64	X = 0.034	X = 0.034	X = 4.91	-
	Y = 14.73	Y = 1.964	Z = 0.034	Y = 0	
	Z = 14.73	Z = 0.034		Z = 7.36	
0.736111111	X = 2.0	X = 0.0035	X = 0.0035	X = 0.5	Wind
	Y = 1.5	Y = 0.2	Z = 0.0035	Y = 0	Tunnel
	Z = 1.5	Z = 0.0035		Z = 0.75	

Boundary conditions were set to *Velocity Inlet* at the inlet, *Outflow* at the outlet, *Wall* for the floor, *Symmetry* for the walls, *Wall* for pollutant walls, and *Velocity Inlet* for the nozzle (spout of the chimney).

Profiles of the flow field physical characteristics were determined at the inlet of the geometry (see Table 2) based on experimental data from a low-speed wind tunnel for a 1:1000-scale model (see [3],[4],[5],[11]). For other model scales, the profiles were modified to keep the trend of curves.

Table 2: Air flow velocity profile and turbulent characteristics profiles in geometry

Title	Equation
Vertical profile of air flow velocity (X-direction)	vx=0.2371·ln(Y+0.00327)+1.3571
Vertical profile of air turbulent intensity	I=−0.0673·ln(Y+0.00327)+0.1405
Vertical profile of turbulent kinetic energy	k=1.5· (vx·I)2
Vertical profile of turbulent dissipation rate	ε=1.225·0.09· (k3)1.4

In Table 2 the parameter v_x represents the air flow velocity in the direction of X-axis [m/s], I is the intensity of turbulence [%], Y is the vertical coordinate of the geometry [m], k is the turbulent kinetic energy [m^2/s^2], and ε is the turbulent dissipation rate [m^2/s^3].

The pollutant source was designed as a nozzle (chimney) of a cuboid shape. Its dimensions and spout surfaces are shown in Table 1. For all model scales, the pollutant velocity $v_{polutant}$ was set at 0.5 [m/s], the intensity of turbulence in the pollutant source at 10 [%], the pollutant mass fraction in the pollutant source at 0.95 [-], and the air mass fraction in the pollutant source at 0.05 [-]. The hydraulic diameter of the pollutant source was set at 3.5 [m] for 1:1-scale model, 0.866 [m] for 1:4.04-scale model, 0.0986 [m] for 1:35.51-scale model, 0.344 [m] for 1:101.821-scale model, and 0.0035 [m] for 1:1000-scale model scale 1:1000.

Three different pollutants were chosen to be tested: helium, methanol and 1, 2-dichlorethane. Helium (ρ = 0.1625 [kg/m^3]) has a lower density than air, i.e., it is lighter than air (ρ = 1.225 [kg/m^3]). Methanol (ρ = 1.43 [kg/m^3]) has approximately the same density as air, i.e., it is approximately of the same weight as air. 1, 2-dichlorethane (ρ = 4.1855 [kg/m^3]) has a greater density than air, i.e., it is heavier than air. Plumes of pollutants lighter than air tend to climb, whereas those heavier than air tend to descend. However, this is not always the case. Pollutant plume vertical movements can be influenced by several other physical factors as demonstrated in the analysis (see Sections 2.4, 2.5 and 2.6).

Analysis of Results by Air Flow Velocity

The aim of this analysis is to compare gas pollutant plume shapes and motions for three different gas pollutants (helium, methanol and 1, 2-dichlorethane) at different values of the air flow velocity v_{air}. The demonstration of the problem was performed with a 1:1-scale three-dimensional geometry representing the real pattern of a simple terrain with a chimney (see Table 1). The referential air flow velocities v_{air} at the level of the chimney spout (pollutant source) were 1 [m/s], 3 [m/s] and 5 [m/s].

The resulting values of the Froude number Fr for each pollutants and air flow velocities are shown in Table 3. Results were calculated using the ANSYS Fluent 13.0 software and were visualized as iso-surfaces of pollutant concentrations or as contours of pollutant concentration fields (see Figure 7 to Figure 12). The contours were plotted in two-dimensional planes of the geometry, sc., the central vertical longitudinal plane, the floor (ground) plane, and the outlet plane.

Table 3: Values of the Froude number for one model scale and three different air flow velocities

Pollutant	Chemical Symbol/Formula	Model Scale [-]	Air Flow Velocity [m/s]	Froude number [-]	Note
Helium	He	1:01	1	0.248	Fr< 1
(gas)		1:01	3	2.23	Fr "/> 1
		1:01	5	6.195	Fr"/> 1
Methanol	CH_3OH	1:01	1	0.028	Fr<< 1
(gas)		1:01	3	0.253	Fr< 1
		1:01	5	0.704	Fr< 1
1,2-dichlorethane (gas)	CH_2ClCH_2Cl	1:01	1	0.01	Fr<< 1
		1:01	3	0.088	Fr<< 1
		1:01	5	0.246	Fr< 1

The following figures show that with increasing air flow velocity v_{air} the pollutant plume vertical movements are reduced. The pollutant plume inclines horizontally at the level of the chimney spout (pollutant source) showing no tendency to climb or descend. This is because the inertial force F_{I-air} increases as the air flow velocity v_{air} increases. Hence, the pollutant plume vertical movements are reduced or totally eliminated.

The air flow velocity also influences the size and shape of the pollutant plume. With increasing air flow velocity v_{air} the pollutant plume tends to be narrower and longer. However, a further increase in the air flow velocity makes the pollutant plume shorter because of greater rate of the pollutant dispersion. The plume range at certain concentration of the pollutant therefore decreases with increasing air flow velocity.

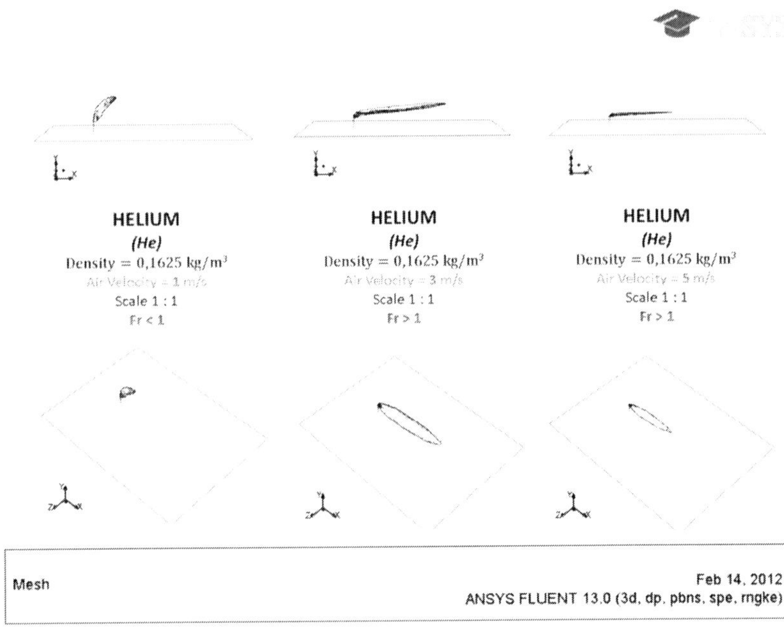

Figure 7: Gas pollutant plume motion analysis by air flow velocity (iso-surfaces of gaseous helium concentration with mass fraction of 0.0001 [-]).

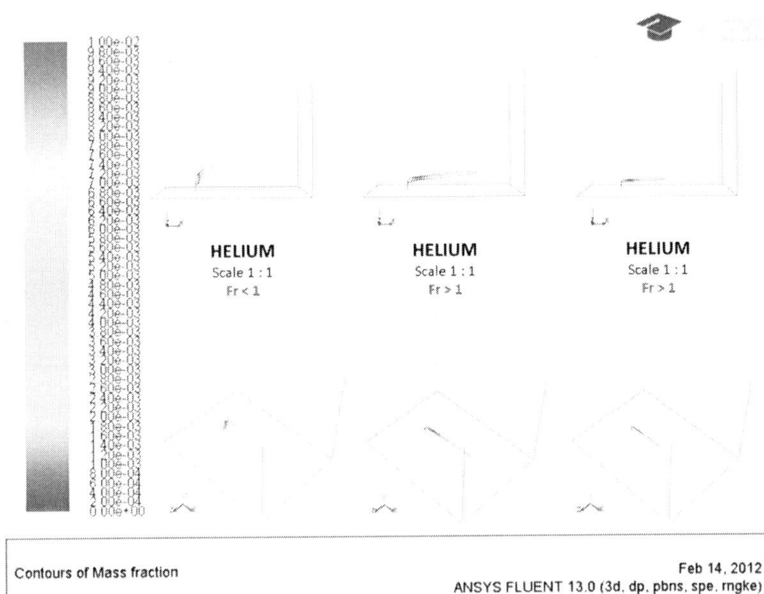

Figure 8: Gas pollutant plume motion analysis by air flow velocity (contours of gaseous helium concentration with mass fraction range of 0–0.01 [-])

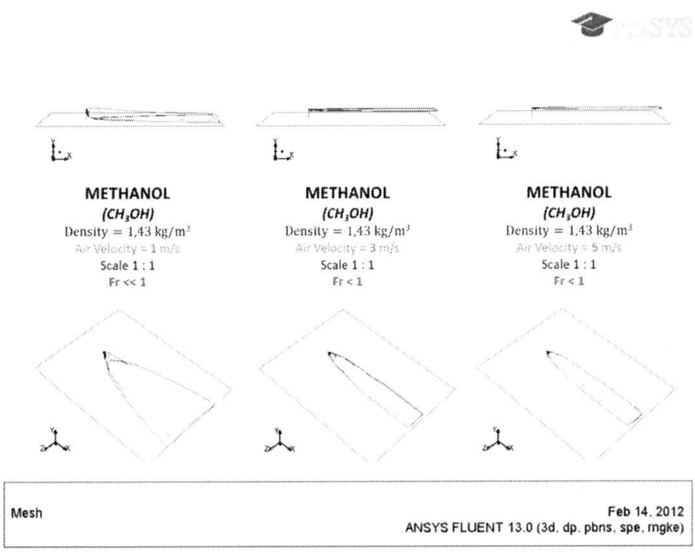

Figure 9: Gas pollutant plume motion analysis by air flow velocity (iso-surfaces of gaseous methanol concentration with mass fraction of 0.0001 [-])

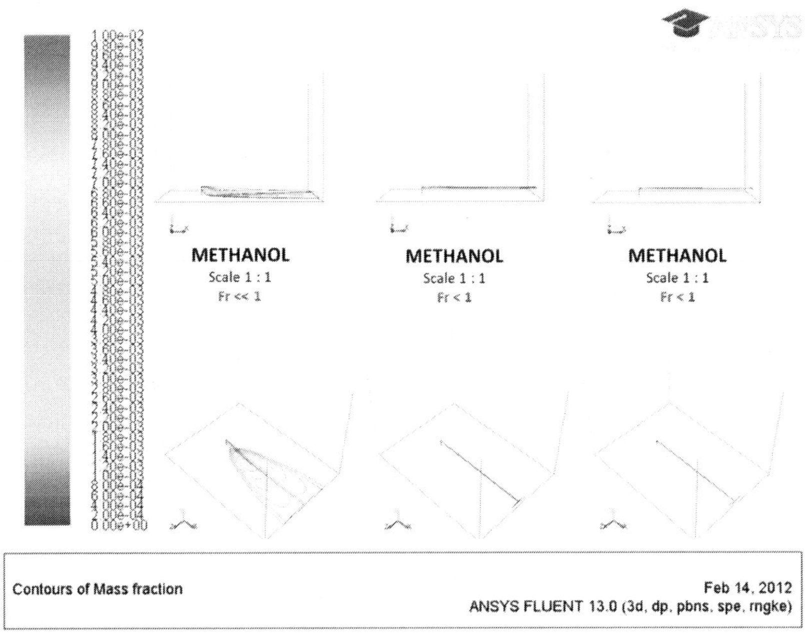

Figure 10: Gas pollutant plume motion analysis by air flow velocity (contours of gaseous methanol concentration with mass fraction range of 0–0.01 [-])

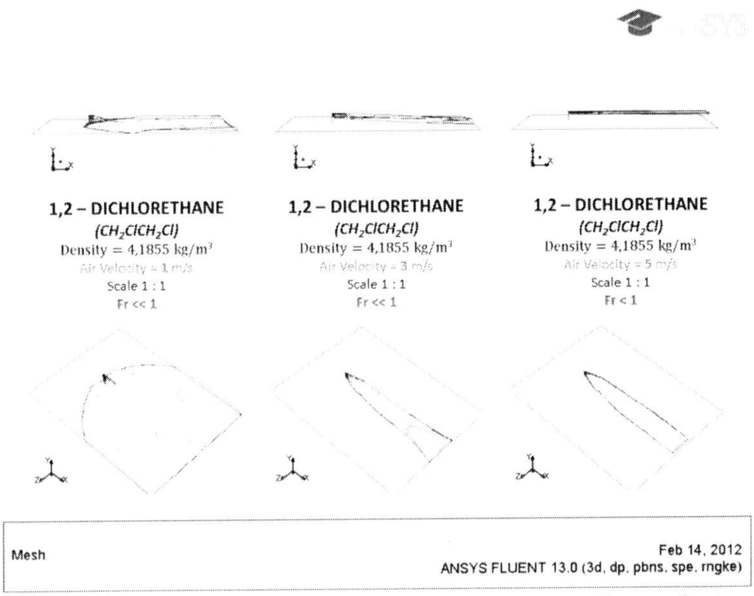

Figure 11: Gas pollutant plume motion analysis by air flow velocity (iso-surfaces of gaseous 1,2-dichlorethane concentration with mass fraction of 0.0001 [-])

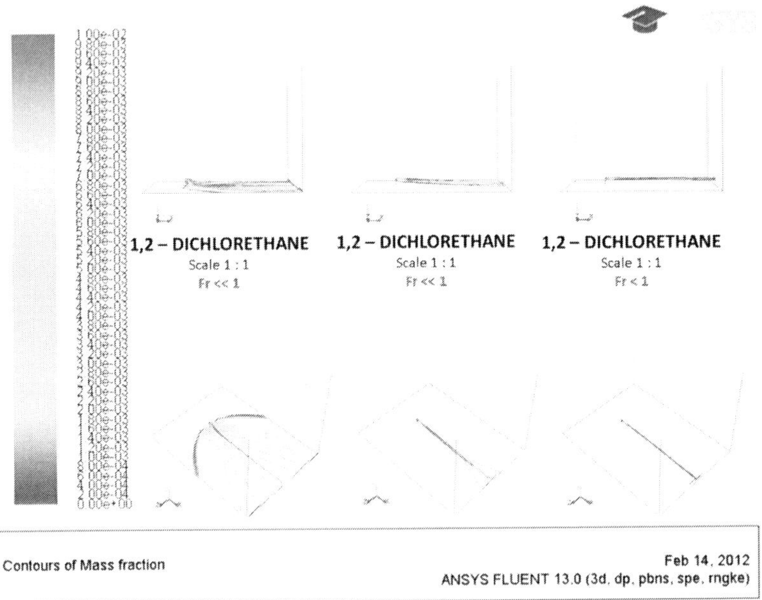

Figure 12: Gas pollutant plume motion analysis by air flow velocity (contours of gaseous 1,2-dichlorethane concentration with mass fraction range of 0–0.01 [-])

Analysis of Results by Pollutant Density

The aim of this analysis is to compare gas pollutant plume shapes and motions for three different gas pollutants (helium, methanol and 1,2-dichlorethane) at different values of their density. Demonstration of the problem was performed with a 1:1-scale three-dimensional geometry representing the real pattern of a simple terrain with a chimney (see Table 1). The referential air flow velocity v_{air} at the level of the chimney spout (pollutant source) was 1 [m/s].

The resulting values of the Froude number Fr for each individual pollutant are shown in Table 4.

Results were calculated using the ANSYS Fluent 13.0 software and were visualized as iso-surfaces of pollutant concentrations or as contours of pollutant concentration fields (see Figure 13 and Figure 14). The contours were plotted in two-dimensional planes of the geometry, sc., the central vertical longitudinal plane, the floor (ground) plane, and the outlet plane.

Table 4: Values of the Froude number for one model scale and one air flow velocity

Pollutant	Chemical Symbol/Formula	Model Scale [-]	Air Flow Velocity [m/s]	Froude number [-]	Note
Helium (gas)	He	1:01	1	0.248	Fr< 1
Methanol (gas)	CH_3OH	1:01	1	0.028	Fr<< 1
1,2-dichlorethane (gas)	CH_2ClCH_2Cl	1:01	1	0.01	Fr<< 1

The figures show that: If the pollutant density $\rho_{pollutant}$ is lower than the air density ρ_{air} the gas pollutant plume tends to climb (for helium see Figure 13 and Figure 14). If the pollutant density $\rho_{pollutant}$ is approximately the same as the air density ρ_{air} the gas pollutant plume neither climbs nor descends (for methanol see Figure 13 and

Figure 14). If the pollutant density $\rho_{pollutant}$ is greater than the air density ρ_{air} the gas pollutant plume tends to descend (for 1,2-dichlorethane see Figure 13 and Figure 14). The range of vertical movements is determined by gravity force $F_{G-pollutant}$ that influences pollutant plume at given conditions. The greater the gravity force $F_{G-pollutant}$ is compared to the inertial force F_{I-air}, the more significant vertical movement of the plume is, i.e., light pollutant plume climbs and heavy pollutant plume descends.

The pollutant density $\rho_{pollutant}$ also influences the pollutant plume dispersion. The greater the pollutant density is, the longer the range of the plume is. At given air flow velocity v_{air}, the plume dispersion of pollutants with a low density is faster and easier than that of pollutants with a greater density.

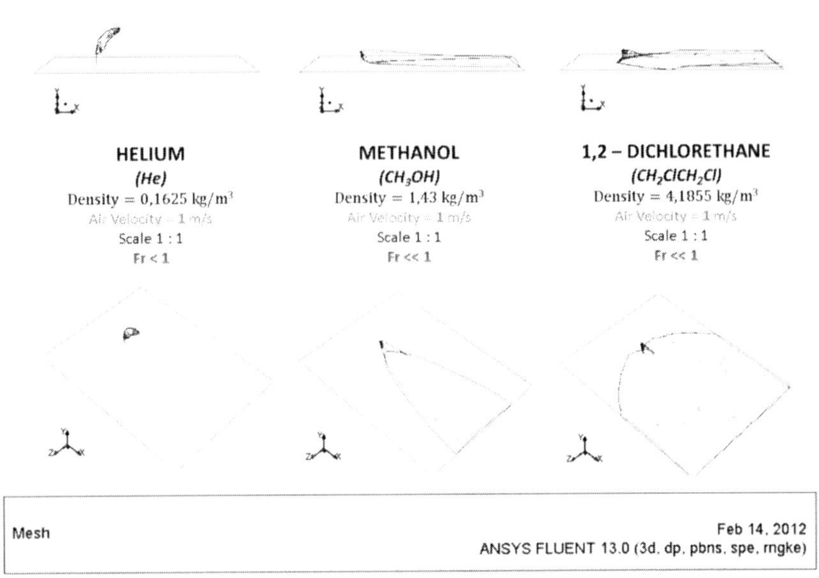

Figure 13: Gas pollutant plume motion analysis by pollutant density (iso-surfaces of gaseous helium, methanol, and 1,2-dichlorethane concentrations with mass fraction of 0.0001 [-]; 1:1-scale model).

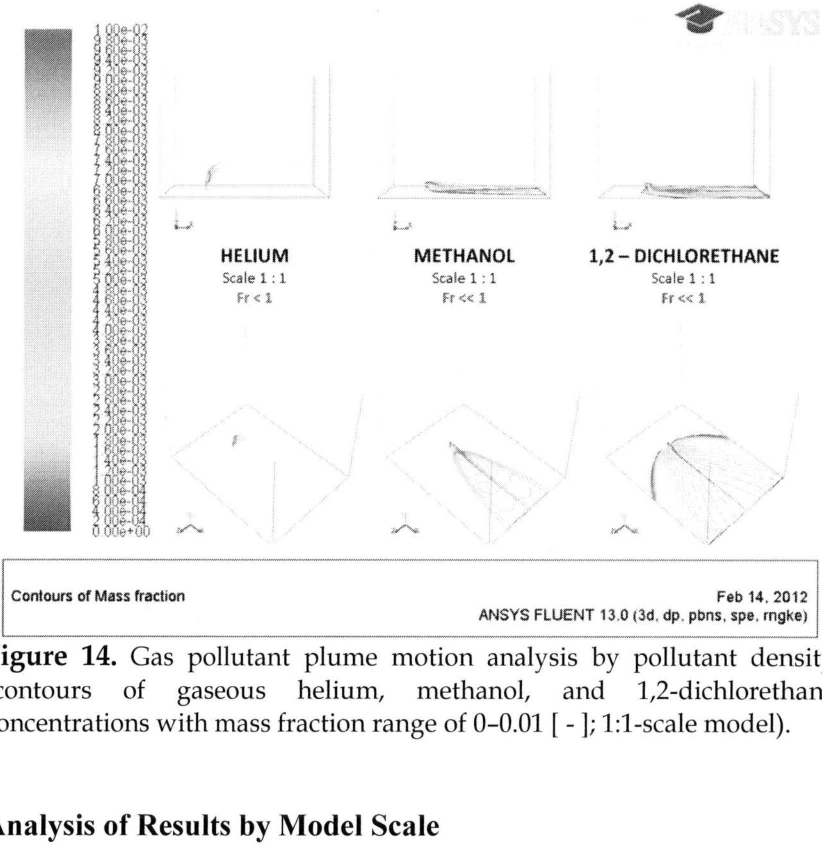

Figure 14. Gas pollutant plume motion analysis by pollutant density (contours of gaseous helium, methanol, and 1,2-dichlorethane concentrations with mass fraction range of 0–0.01 [-]; 1:1-scale model).

Analysis of Results by Model Scale

The aim of this analysis is to compare gas pollutant plume shapes and motions for three different gas pollutants (helium, methanol, and 1,2-dichlorethane) at different model scales. The demonstration of the problem was performed with a three-dimensional geometry at three selected scales (see Table 1) for each of the three pollutants. The first 1:1-scale model represents the real pattern of a simple terrain with a chimney (pollutant source) where Fr< 1, i.e., the gravity force $F_{G-pollutant}$ is greater than the inertial force F_{I-air}. The second model (scaled at 1:4.04, 1:35.51, and 101.82, respectively) represents the state when Fr = 1, i.e., the gravity force $F_{G-pollutant}$ equals the inertial force F_{I-air}. The third 1:1000-scale model represents the gauging section of a low-speed wind tunnel with a nozzle (pollutant source) on the floor where Fr> 1, i.e., the inertial force F_{I-air} is greater than the gravity force $F_{G-pollutant}$. The referential air flow velocity v_{air} at the level of the chimney spout (pollutant source) was 1 [m/s].

The resulting values of the Froude number Fr for each pollutant and model scale are shown in Table 5. Results were calculated using the ANSYS Fluent 13.0 software and were visualized as iso-surfaces of pollutant concentration or as contours of pollutant concentration fields (see Figure 15 to Figure 20). Contours were plotted in two-dimensional planes of the geometry, sc. the central vertical longitudinal plane, the floor (ground) plane, and the outlet plane.

Table 5: Values of the Froude number for different model scales and one air flow velocity

Pollutant	Chemical Symbol/Formula	Model Scale [-]	Air Flow Velocity [m/s]	Froude number [-]	Note
Helium	He	1:01	1	0.248	Fr< 1
(gas)		01:04.0	1	1	Fr = 1
		0.736111111	1	247.806	Fr "/>"/ > 1
Methanol	CH_3OH	1:01	1	0.028	Fr<< 1
(gas)		01:35.5	1	1	Fr = 1
		0.736111111	1	28.16	Fr "/>"/ > 1
1,2-dichlorethane(gas)	CH_2ClCH_2Cl	1:01	1	0.01	Fr<< 1
		02:41.8	1	1	Fr = 1
		0.736111111	1	9.822	Fr "/> 1

The figures (see Figure 15 to Figure 20) show that: If the model scale changes and all other characteristics remain unchanged, the inertial and gravity forces and their ratio change too. Therefore, the size, shape and inclination of the pollutant plume change.

According to Equation (11) in Section 1.5, the inertial force is proportional to the square of the model scale. According to Equation (12) in Section 1.5, the gravity force is proportional to the third power of the model scale. Therefore, the change in the gravity force due to the change of the model scale is considerably greater than the change in the inertial force. The lower the model scale is, the greater the dominance of inertial forces is compared to gravity forces), and vice versa.

Also, the greater the pollutant density, the lower the model scale is if Fr= 1, i.e., the inertial and gravity forces are equal.

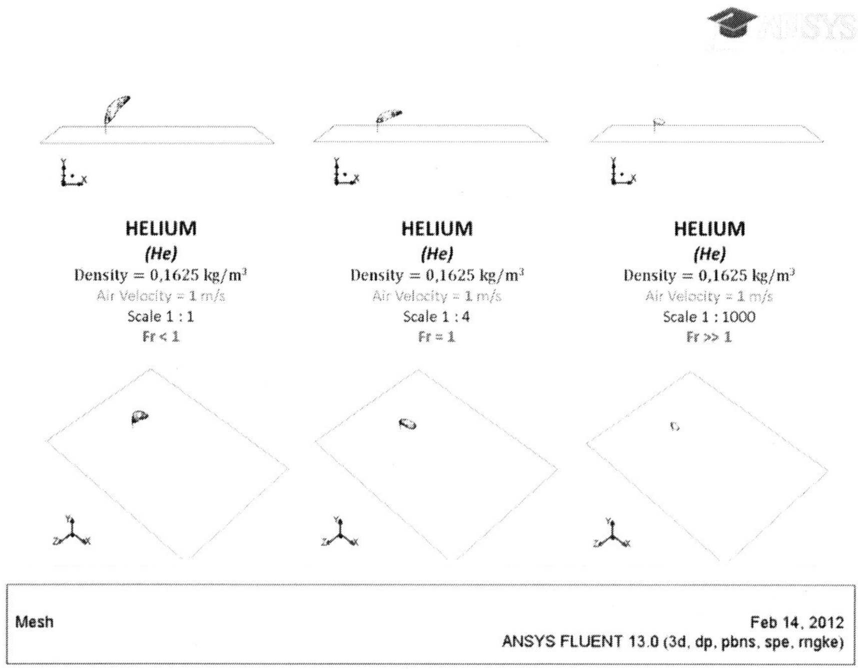

Figure 15: Gas pollutant plume motion analysis by model scale (iso-surfaces of gaseous helium concentration with mass fraction of 0.0001 [-] for three different model scales).

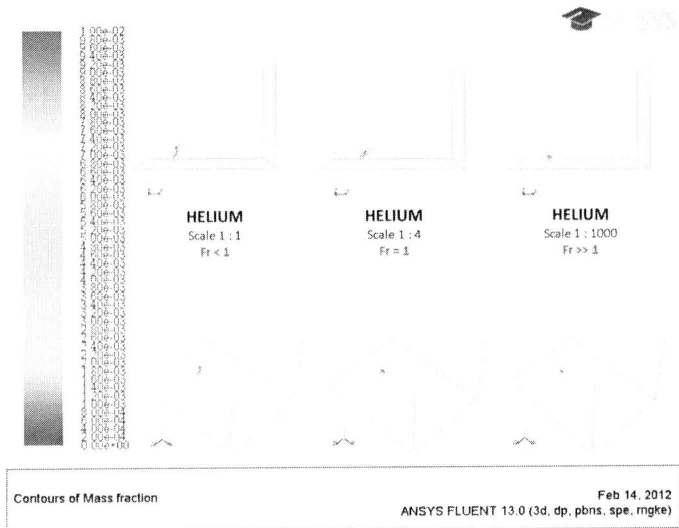

Figure 16: Gas pollutant plume motion analysis by model scale (contours of gaseous helium concentration with mass fraction range of 0–0.01 [-] for three different model scales).

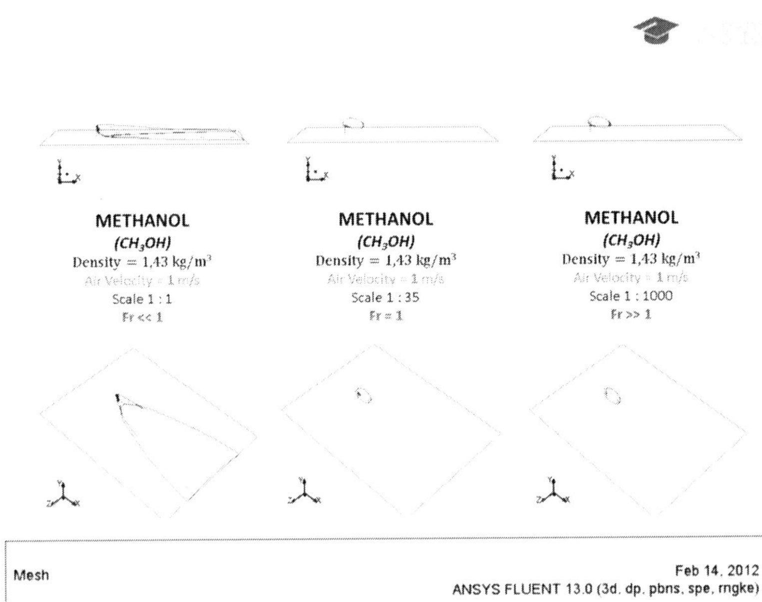

Figure 17: Gas pollutant plume motion analysis by model scale (iso-surfaces of gaseous methanol concentration with mass fraction of 0.0001 [-] for three different model scales).

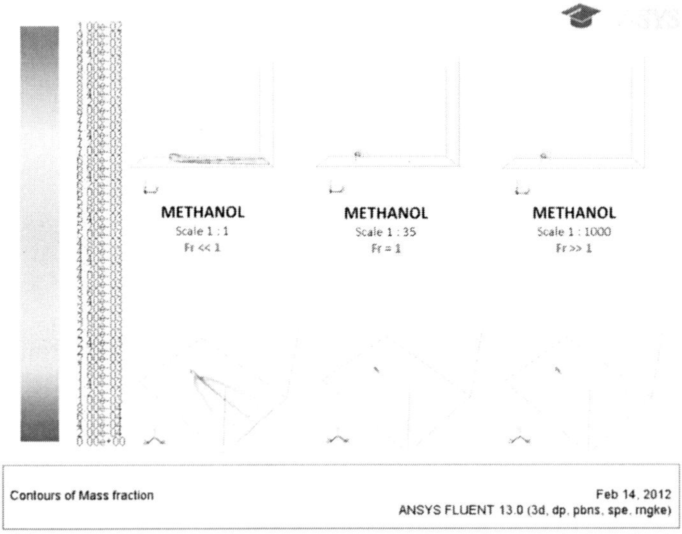

Figure 18: Gas pollutant plume motion analysis by model scale (contours of gaseous methanol concentration with mass fraction range of 0–0.01 [-] for three different model scales).

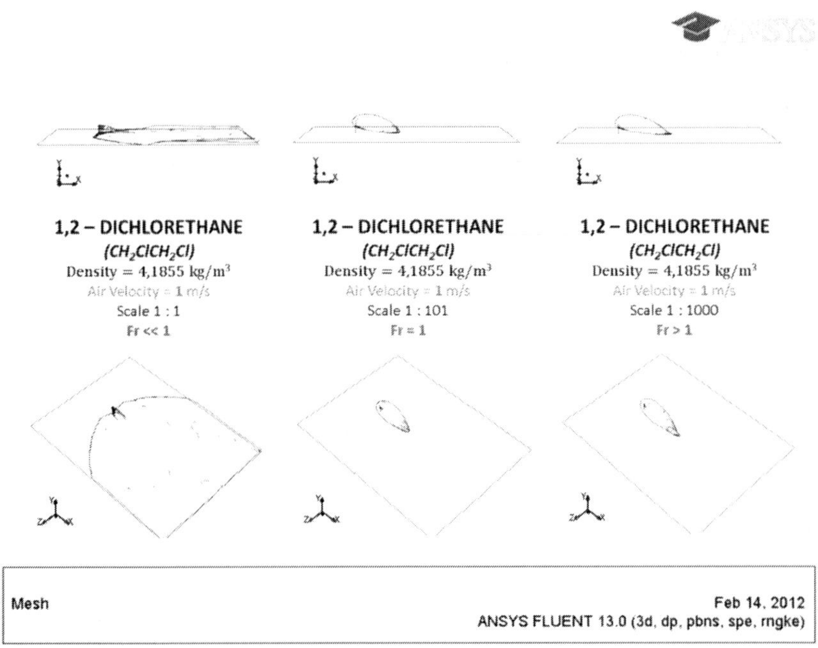

Figure 19: Gas pollutant plume motion analysis by model scale (iso-surfaces of gaseous 1,2-dichlorethane concentration with mass fraction of 0.0001 [-] for three different model scales).

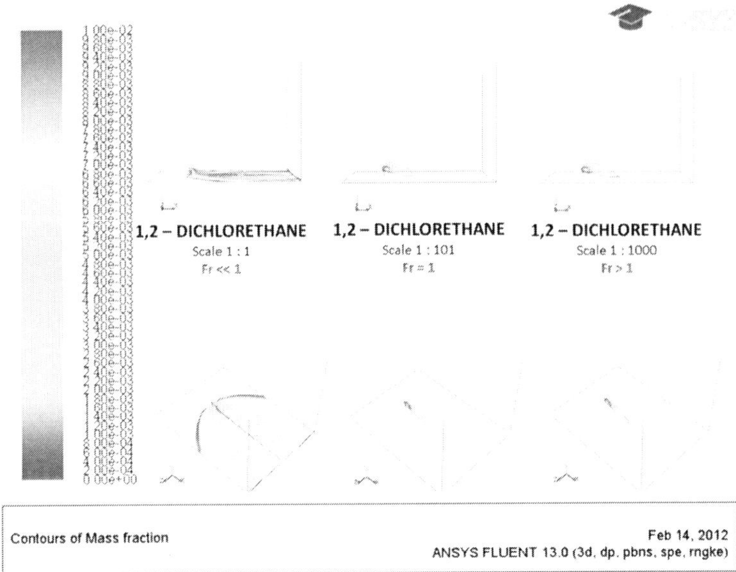

Figure 20: Gas pollutant plume motion analysis by model scale (contours of gaseous 1,2-dichlorethane concentration with mass fraction range 0–0.01 [-] for three different model scales).

CONCLUSION

The aim of the analyses was to lay down principles for physical and mathematical modeling of gas pollutant plume motion and dispersion in real atmospheric conditions. The influences of the air flow velocity, pollutant's density, and model scale on pollutant plume size, shape, and inclination were investigated.

The Froude number was chosen as a criterion of physical similarity for the pollutant plume behavior in the atmosphere. Basic mathematical rules and principles were formulated upon study of available fluid mechanics literature (see [7],[8],[9],[10],[11]). All mathematical and physical assumptions were next verified by numerical simulation using the ANSYS Fluent 13.0 software. Air flow field was modeled using the RNG k−ε model of turbulence, the gas pollutant motion was modeled using the Species Transport Model, both in the same three-dimensional geometry consisting of 569 490 grid cells. Turbulent characteristics were defined using RANS approach. No additional dispersion model was applied.

Object of modeling was gauging section of the low speed wind tunnel (for model scale of 1:1000) or big real terrain (for model scale of 1:1) with a pollutant source in form of nozzle (or chimney, respectively) situated on the section floor (ground). The gauging section with the nozzle represented a chimney in a simple, flat terrain. The chimney was considered to be a pollutant source for three different gas pollutants (helium, methanol, and 1,2-dichlorethane). The numerical simulation was performed for five model scales, three gas pollutants with different densities, and three different air flow velocities. The simulations were steady (time-independent) with the accuracy of 0.0001 (criterion of convergence). Final results were visualized as iso-surfaces of pollutant concentrations and contours of pollutant concentration fields with the concentration limit value of 0,001. The contours were plotted in two-dimensional planes of the geometry, sc. the central vertical longitudinal plane, the floor (ground) plane and the outlet plane. The numerical model had been verified by an experiment performed in a low-speed wind tunnel (see [3],[4],[5]).

The following principles based on the results of the Froude number analysis of pollutant plume motion and dispersion in real atmosphere can be defined:

1. The greater the air flow velocity is, the greater the inertial forces are. These forces influence pollutant plume and reduce its vertical motions (inclination). With increasing air flow velocity the pollutant plume inclines horizontally at the level of the chimney spout (pollutant source), but with further increase in the air flow velocity it becomes narrower and shorter (see Section 2.4).
2. The greater the difference between pollutant density and air density is, the more significant the tendency towards vertical movements (climbing or descending) of the plume is. The plume of pollutant with lower density than air tends to climb, whereas the plume of pollutant with greater density than air tends to descend. The density of the pollutant also influences the pollutant plume dispersion. The greater the pollutant density is, the longer the range of the plume is. At given air flow velocity v_{air}, the plume dispersion of pollutants with a low

density is faster and easier than that of pollutants with a greater density (see Section 2.5).

3. If the model scale changes and all other characteristics remain unchanged, the inertial and gravity forces and their ratio change too. Therefore, the size, shape and inclination of the pollutant plume change. The inertial force is proportional to the square of the model scale, whereas the gravity force is proportional to the third power of the model scale (see Section 1.5). The change in the gravity force due to the change of model scale is considerably greater than the change in the inertial force. The lower the model scale is, the greater the dominance of inertial forces is compared to gravity forces. Also, the greater the pollutant density is, the lower the model scale is if Fr= 1, i.e., the inertial and gravity forces are equal (see Section 2.6).

From the above it follows that if investigators want to respect and follow the basics of physical phenomena, they must consider criteria of physical similarity very carefully, in particular criteria of dynamic similarity (see Section 1.1). Some physical phenomena, however, cannot be modeled in any model scale but the original one without changing the basis of the phenomena (see Sections 1.5 and 2.6).

This analysis is intended for those who are interested in gas pollutant plume motion in the atmosphere and in theory of physical similarity. The conclusions of the analysis can be used for further experiment design works or for checking results of mathematical modeling.

ACKNOWLEDGEMENT

The author acknowledges the financial support of the SPII 1a10 45/07 project of the Ministry of the Environment of the Czech Republic.

REFERENCES

1. M. Kozubkova, 2008Modeling of Fluid Flow (in Czech), FLUENT, CFX. Ostrava: VSB- Technical University of Ostrava. 153 p.

2. M. Bojko, 2008Guide for Training of Flow Modeling- FLUENT (in Czech). Ostrava: VSB- Technical University of Ostrava. 141 p.
3. S. Civis, Z. Zelinger, M. Strizik, Z. Janour, 2001Simulation of Air Pollution in a Wind Tunnel. Spectroscopy from Space. Dordrecht: Kluwer Academic. 275299
4. Z. Zelinger, M. Strizik, P. Kubat, Z. Janour, P. Berger, A. Cerny, P. Engst, 2004Laser Remote Sensing and Photoacoustic Spectrometry Applied in Air Pollution Investigation. Opt. Lasers Eng. 42. 403412
5. Z. Zelinger, M. Strizik, P. Kubat, S. Civis, E. Grigorova, R. Janeckova, O. Zavila, V. Nevrly, L. Herecova, S. Bailleux, V. Horka, M. Ferus, J. Skrinsky, M. Kozubkova, S. Drabkova, Z. Janour, 2009Dispersion of Light and Heavy Pollutants in Urban Scale Models: CO2 Laser Photoacoustic Studies. Applied spectroscopy. Society for Applied Spectroscopy. 430436
6. Benson, Tom2010Wind Tunnel Index. Available: http://www.grc.nasa.gov/WWW/k-12/airplane/shortt.html.Accessed 2012 March 4.
7. M. Carnogurska, M. Prihoda, 2011Application of Three-dimensional Analysis for Modelling Phenomena in the field of Power Engineering (in Slovak). Kosice: Technical University of Kosice. 214 p.
8. Incropera F.P, Dewitt D.P, Bergman T.L, Lavine A.S2007Fundamentals of Heat and Mass Transfer. New York: John Wiley & Sons. 997 p.
9. Shaughnessy E.J, Katz M.I, Schaffer J.P2005Introduction to Fluid Mechanics. New York: Oxford University Press. 1118 p.
10. S. Drabkova, P. Platos, 2003Numerical Simulation as a Tool for the Solution and Understanding of Practical Air Pollution Problems. Proc. of the Conference on Modelling Fluid Flow (CMFF'03). Budapest: Budapest University of Technology and Economics. 501506
11. Stull B.R1994An Introduction to Boundary Layer Meteorology. Dordrecht: Kluwer Academic Publisher. 666 p.

CITATION

OndrejZavila (2012). Physical Modeling of Gas Pollutant Motion in the Atmosphere, Advances in Modeling of Fluid Dynamics, Dr. Chaoqun Liu (Ed.), ISBN: 978-953-51-0834-4, InTech, DOI: 10.5772/48405.

CHAPTER 8

Applications of CFD in Natural Gas Processing and Transportation

Majid Abedinzadegan Abdi[1], *Esam Jassim, Mohammad Haghighi and Yuri Muzychka*

[1]Memorial University of Newfoundland, St. John's, Newfoundland and Labrador, Canada

INTRODUCTION

In this chapter, two examples of CFD applications in natural gas processing and transportation are presented. A commercial software package (Fluent) was used in these studies. The purpose of the studies is briefly discussed, the methodology outlined and boundary conditions and problem specifications are concisely described for each case. The results of investigations and comparisons with experimental tests and literature data are presented to demonstrate how CFD can be applied to practical situations.

FLOW OF REAL GAS IN SUPERSONIC NOZZLES

The demand for natural gas has encouraged the energy industry toward the discovery of remote offshore reservoirs. Consequently new technologies have to be developed to efficiently produce and transport natural gas to consumption centers. Common design challenges in all gas processing methods for offshore applications are the compactness and reliability of process equipment. Supersonic nozzles have been introduced as an alternative to treat natural gas for offshore applications and to meet the offshore requirements (Hengwei et al. 2005, Alfyorov et al. 2005,Okimoto et al., 2002, Karimi & Abedinzadegan Abdi, 2006, Brouwer & Epsom, 2003). In a supersonic separator the gas temperature is lowered based on the principle of gas expansion where no refrigerant is needed. The compact design of supersonic nozzles is a major advantage over traditional means of natural gas treating technologies particularly for offshore applications. The gas speed in this device is very high preventing fouling or deposition of solids and ice. Refrigeration is self-induced therefore no heat is transferred through the walls and unlike external refrigeration systems, no inhibitor injection and inhibitor recovery system are necessary. Intensive water dew points down to -50 to -60 ^{o}C can be achieved without any cryogenic cooling or use of solid adsorption techniques.

Problem Description

Application of CFD technique is demonstrated to predict the behaviour of high pressure natural gas flowing through supersonic nozzles. Supersonic nozzles were selected as it was noticed that there was a potential for these nozzles for applications in natural gas processing industries and very few simulation analysis had been published in the open literature. The nozzle considered here is a de Laval geometry composed of two sections: the convergent section (subsonic zone) and the divergent section (supersonic zone). However, we also address two other de Laval modified geometries, which are of interest in solid/liquid particles separation; namely throat section (critical zone) with extended constant area throat, and throat section with extended U-shape throat.

The function of the convergent part is to keep the flow uniform and parallel as well as to accelerate the gas. Within the converging section leading to the throat area, the gas is accelerated so that the sonic velocity is reached at the throat and the convergent curvature keeps the gradient in velocity of the flow uniform. In practical conditions, in order to get the sonic speed at the throat, it is required that the inlet diameter is kept larger than 5 of the throat diameter (Man et al., 1997) although in some cases the ratio of inlet to throat diameter is apparently less (Arina, 2004).

When the gas reaches the throat, the divergent part of the nozzle can further accelerate the flow depending on the outlet condition. This results in a decrease in pressure and temperature as well as increase in gas velocity. It is likely that under certain conditions the flow cannot expand isentropically to the exit pressure; therefore, an irreversible discontinuity, called a shockwave, can occur.

The shockwave is an abrupt disturbance that causes discontinuous and irreversible changes in fluid properties, such as speed (changing from supersonic to subsonic), pressure, temperature, and density. As a result of the gradients in temperature and velocity that are created by the shock, heat is transferred and energy is dissipated within the gas. These processes are thermodynamically irreversible. As the shock thickness is very small, the cross sectional areas at the upstream and downstream of the shock are considered equal and the energy or heat loss is negligible. The shock can also interact with the boundary layer and this can delay the transition from supersonic flow to subsonic flow even further. The increase of pressure across a shock is an indication of the shock strength that can lead to a sound wave considered as a shockwave of minimum strength.

Basis of CFD Simulation

The geometry was modeled using two-dimensional axisymmetric grids. The total pressure and temperature for fully developed turbulent flow were imposed at the nozzle inlet, and no-slip condition was applied at wall boundaries. At the exit plane, ambient pressure and temperature were identified. CFD calculations were carried out using SIMPLEC algorithm and the central differencing scheme.

For turbulent flow model, the k-ε model was used here due to its frequent use for industrial applications, its relative accuracy, and its incorporation in most commercial CFD codes (Pope, 2000).

Results and Discussions

De Laval Nozzles

Since most of published research has been concerned with the Laval nozzle, we validated our results by applying the numerical technique for such geometry and compared our results with the most recent available data (Arina, 2004; and Molleson & Stasenko, 2005) before proceeding and applying it to the modified nozzle systems.

Molleson and Stasenko (2005) performed their investigation for a nozzle whose geometry is shown inFigure 1-a. Their working fluid was methane at 70 bar inlet stagnation pressure and 250 K inlet stagnation temperature while the value of exit Mach number according to their supersonic exit radii was 1.2. The SRK EOS model was used in their study. We used the same geometry and conditions in comparison of the sonic condition and in studying of the effect of a real gas model on the sonic position; MBWR was used as the thermodynamic model.

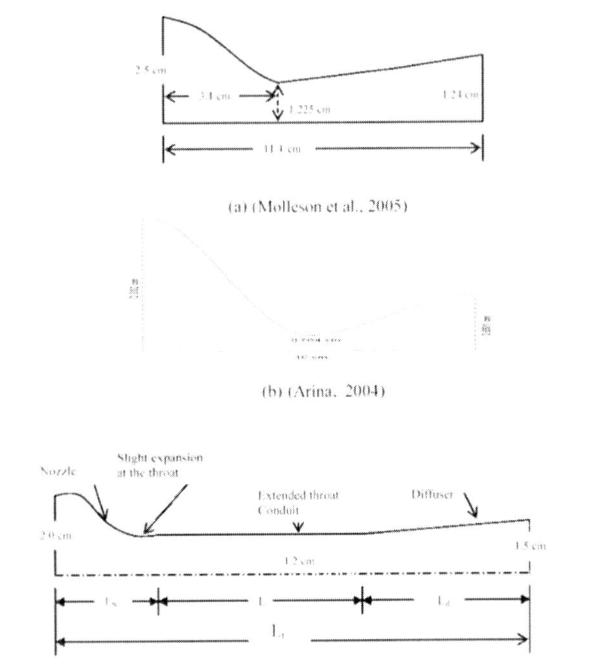

Figure 1: Nozzle geometries studied in the research.

Figure 2 shows the variation of Mach number with position in real case and the comparison with results obtained from the work of Molleson and Stasenko (2005). One can see that choke (sonic velocity, M=1) occurs at the throat regardless of the EOS used. Also, our results are in very good agreement with their results. The second comparison was performed to validate our simulation on capturing shockwave position. The comparison is done with recent available data (Arina, 2004). The geometry used in the comparison, shown in Figure 1-b, is adopted from Arina's work (2004). The working fluid was CO_2. The dimensions of the assumed Laval-nozzle are:

$$A(x) = 2.5 + 3\left(\frac{x}{x_{th}} - 1.5\right)\left(\frac{x}{x_{th}}\right)^2 \quad \text{for} \quad x \le x_{th}, \tag{1}$$

$$A(x) = 3.5 - \frac{x}{x_{th}}\left(6 - 4.5\frac{x}{x_{th}} + \left(\frac{x}{x_{th}}\right)^2\right) \quad \text{for} \quad x \ge x_{th}. \tag{2}$$

Where, A_{throat}= 1 cm^2, length= 10 *cm* and the throat placed at x_{th}= 5 *cm*.

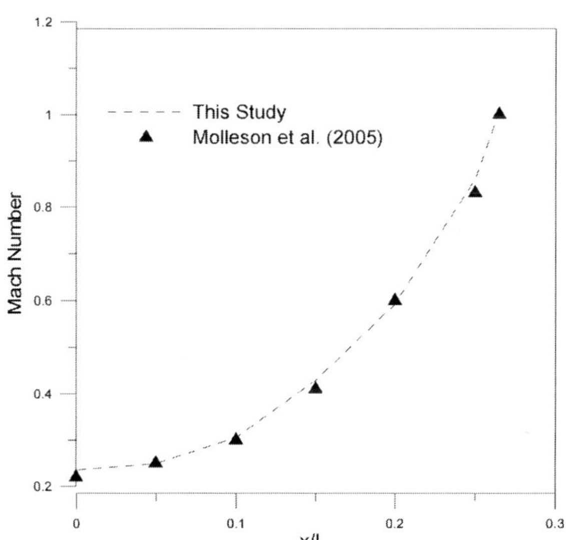

Figure 2: Comparison of Mach numbers upstream of choke region in the Laval nozzle.

Since Arina's simulation was performed near CO_2 condensation conditions, for which $T= 1.001T_c$ and = $_c$, we compared our results for a perfect gas case as the FLUENT real gas basis could not predict multiphase conditions. The exit pressure was 83% of the inlet pressure. Our numerical results displayed the same behaviour when similar conditions and working fluids were applied as seen in Figure 3.

Real Gas vs. Ideal Gas Assumption

The significance of using real gas models can be more clearly shown when comparison of the location of the shockwave within the Laval nozzle is made for two different gases: methane and nitrogen. At high pressures, the former compressibility factor significantly changes whereas the compressibility for the latter has almost the equivalent value of perfect gases.

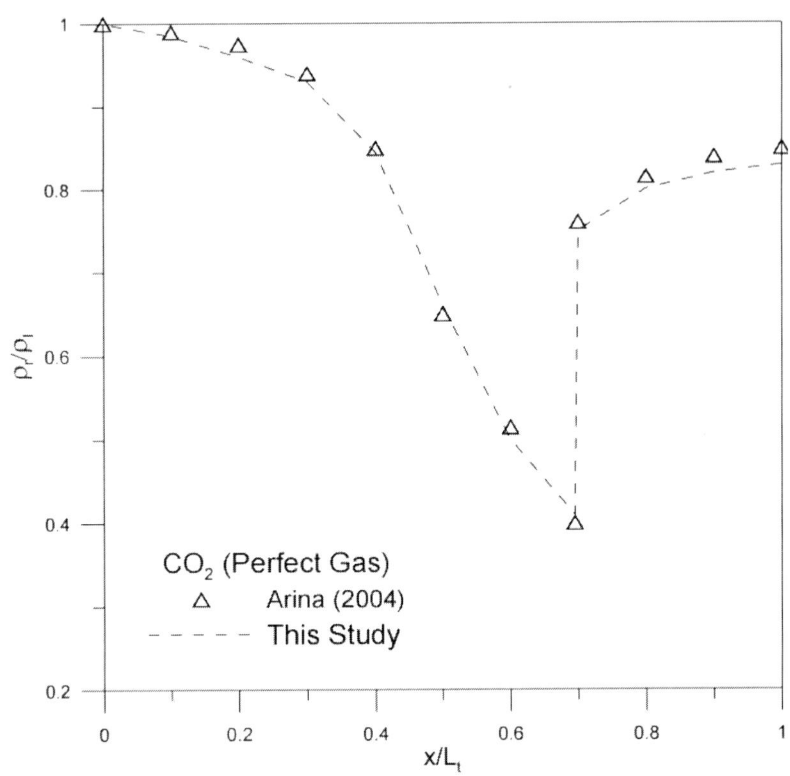

Figure 3: Comparison of density ratios in the Laval nozzle under perfect gas conditions.

Shockwave Location

The real gas model predicts the shockwave location earlier than the ideal gas model for both gases. However, the differential distance between the ideal and real shock positions is different. In fact, when the ideal gas model is used, shock occurs earlier for nitrogen, see Figure 4-a. On the contrary, the real gas model predicts an opposite behaviour as shown in Figure 4-b. This example proves very clearly that ignoring the real gas effects can obviously lead to misleading results.

Real Gas Effects for a Different Configuration

The new configuration of the nozzle system designed for natural gas application consists of three different parts: an inlet nozzle (converging part ending to a throat and a slight expansion), diffuser (diverging part, gas final expansion and exit), and a conduit with constant area between these two parts. This latter part does not exist in conventional Laval nozzles where the diffuser or diverging part starts right after the throat and continues uniformly right up to the exit point. The description of the new system is shown in Figure 1-c.

Boundary conditions were chosen in such a way that the inlet pressure was predicted. Hence, we chose mass flow rate and temperature as the inlet boundary conditions while pressure and temperature were chosen for outlet boundary conditions. The working fluid was pure methane, mass flow rate was 430 kg/minute, stagnation temperature at the inlet and outlet were 293 and 280 K, respectively and the stagnation pressure at the outlet was assumed 7 MPa. The stagnation pressure at the inlet was to be predicted. The results of simulation are discussed as follows:

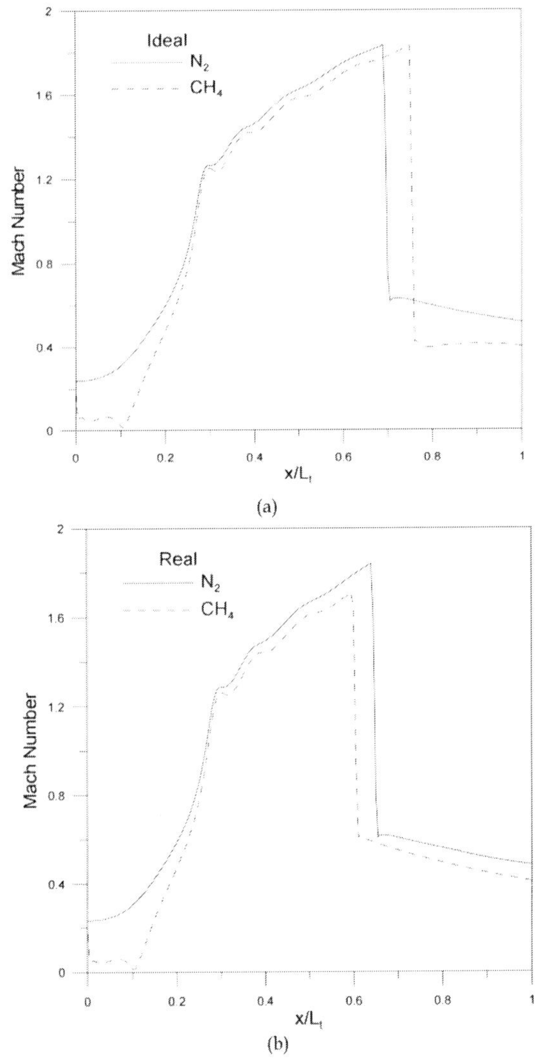

Figure 4: Comparison of shock position for nitrogen and methane under (a) ideal and (b) real gas conditions.

Density: By looking at any fluid textbook, one can see that the conservation of momentum equation directly or indirectly contains density terms in each component. Thus, flow structure is severely affected by any deviation in density calculation. To realize how this deviation will affect the predictions in the nozzle, a graph of the density ratio (real/ideal) along the nozzle system is plotted (see Figure 5). It is evident how erroneous the results might get if the perfect gas model is used, particularly in the vicinity of the shockwave. A large spike of density variations is seen close to the shockwave.

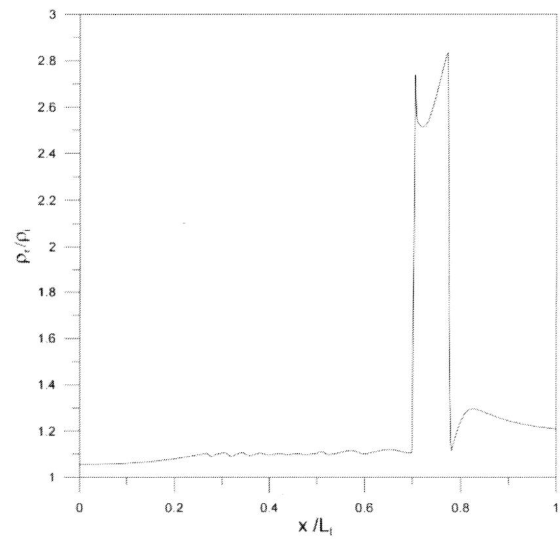

Figure 5: Centerline density ratio for real gas (methane) simulation in an extended throat nozzle.

Inlet Pressure: As mentioned earlier, our numerical solution predicts the inlet pressure for the given mass flow rate, outlet pressure and outlet temperature. Figure 6 represents static pressure distribution at the axis. One can see that a significant difference between the real and perfect gas at the inlet pressure is obtained. The real gas model predicts lower required inlet pressure for a given mass flow rate. Thus, better pressure recovery may be obtained. The real gas simulation predicts pressure recoveries in excess of 10% over those predicted by the perfect gas model. Also, the difference between real and ideal static pressures forces the calculation of total pressure to diverge. The errors in evaluating the total pressure and temperature can lead to incorrect predictions for friction loss, work and heat transfer. The differences between the predictions of ideal and real gas models for total pressure and total temperature are considerable. These discrepancies can result in incorrect values for friction losses and other calculated parameters.

Temperature. Static temperature decreases during the isentropic expansion process. Figure 7illustrates the longitudinal variation of static temperature along the insulated wall. It is clearly shown that the temperature reduction in the real gas case is larger than the ideal case. Thus, ideal gas simulation can lead to erroneous results in predicting the potential phase change.

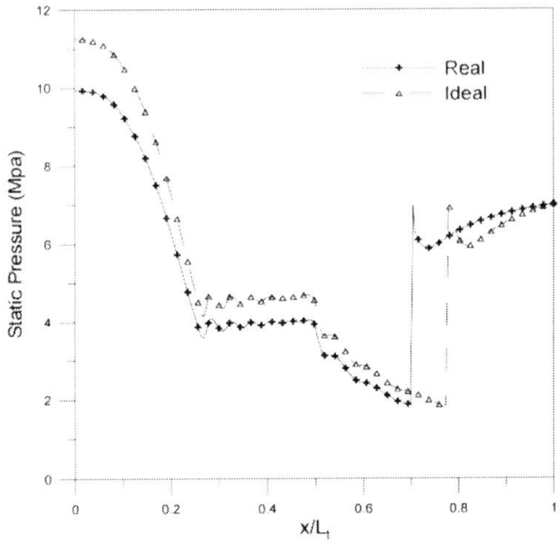

Figure 6: Centerline static pressure for real gas (methane) simulation in an extended throat nozzle.

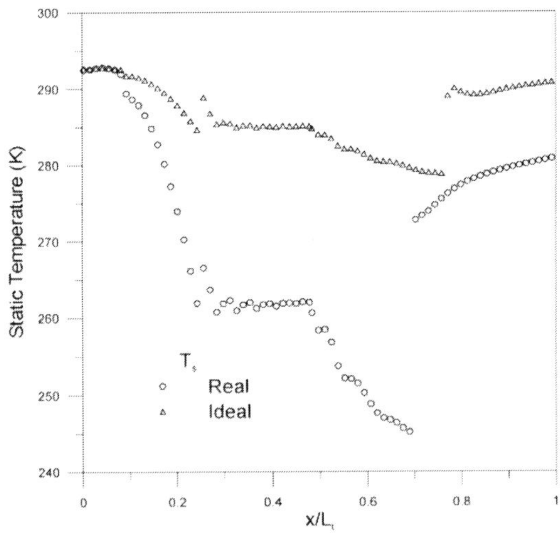

Figure 7: Wall static temperature distribution for real gas (methane) simulation in an extended throat nozzle.

Theoretically, stagnation temperature should retain its value across the shock when the nozzle is insulated. Such conclusion is true when perfect gas law is used as the thermodynamic model. However, the real gas model predicts different conditions in which the stagnation temperature may vary. Stagnation temperature in the real gas case varies across the shock because the specific heat is now varied across the shock as well. Thus, for adiabatic process;

$$\int_{T_{01}}^{T_{02}} Cp(T_0).dT_0 = 0 \tag{3}$$

This means that $T_{02} \neq T_{01}$ across the shockwave.

As a conclusion of modelling of supersonic nozzles using real gas models and discussion of the effect of real gas on the flow of natural gas through these nozzles, the choice of thermodynamic model can substantially affect the modelling results. This includes the position of shockwave, the fluid properties and conditions after the shock, and the work/heat transferred across the system.

De Laval Nozzles with Extended Straight Throat

In this section the influence of geometry on the flow of natural gas through the supersonic nozzle is presented. As mentioned in Section 2.3.1, the nozzle is composed of three sections: the convergent section (subsonic zone), the extended throat section (critical zone) and the divergent section (supersonic zone).

The governing equations are derived from the basic conservation laws including mass (continuity), momentum, and the first and second laws of thermodynamics and the use of a quite suitable real gas Equation of State (EOS).

Geometry Influence

Choke location: The flow reaches sonic condition at the throat, expands in the slight diverging/expansion section of the nozzle (~ 0.04 m), passes through the constant -cross -section area, and finally moves across the diffuser within which the shock occurs. The flow downstream of the shock is subsonic. Thus a pressure recovery occurs. The base Fluent simulation cannot predict any phase change therefore it is assumed that no condensation will occur as the gas passes through the nozzle. Gas dynamics parameters for real and ideal flow are presented in Figure 8. The Figure indicates that no sensible variation is predicted in longitudinal Mach number, especially in the inlet/converging part. The vertical line x=x* confirms that, for both cases, the Mach number value of unity is obtained at the critical cross section of the nozzle. This conclusion agreed very well with the predictions in recent studies (Arina, 2004, Molleson & Stasenko, 2005, and Drikakis & Tsangaris, 1993). It can, therefore, be concluded that the sonic position always occurs at the throat and is independent of the nozzle geometry and gas thermodynamic model.

Shock position: A shock wave occurs in the diffuser part of the system that leads to a change in the flow from supersonic to subsonic. The real gas model predicts the shock location earlier as concluded in earlier section. Arina (2004) compared several real gas EOS with the corresponding perfect gas model. He concluded that all the real gas models predict a similar flow in the convergent part, while the shock position is slightly varying from one model to another. However, his conclusion was built on a Laval nozzle and inert gases (air) which behave almost ideally even at high pressures.

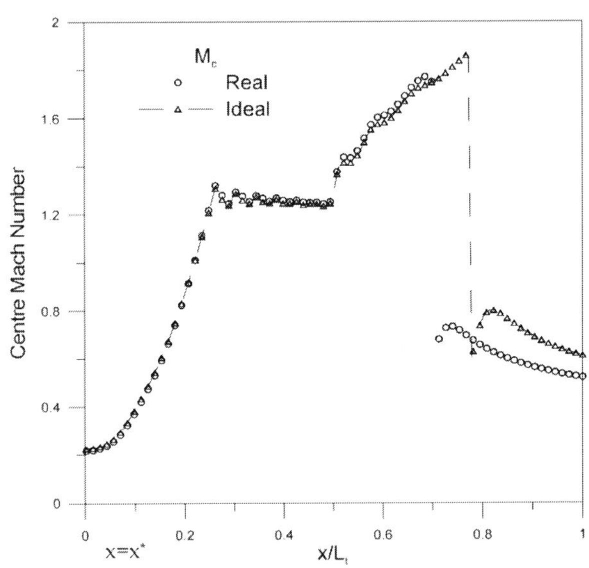

Figure 8: Centerline Mach number for real vs. ideal methane simulation in an extended throat nozzle.

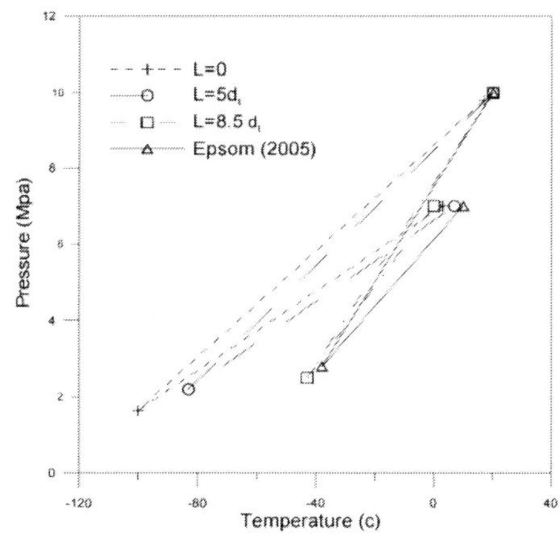

Figure 9: Pressure-temperature chart for real gas (methane) simulation in an extended throat nozzle.

It can also be concluded that thermodynamic models and the system's geometry play the most significant roles in predicting the shock position. Hence, the more accurate EOS would result in a closer prediction of the shock location and behaviour.

Length of Constant Cross Section Area Conduit: In order to separate liquid particles from natural gas and place instruments to control the shock wave location, Twister Inc. (Epsom, 2005) uses an extended constant area throat within the area just before the diffuser section of the Laval nozzle. Although the exact dimension and geometry of the Twister's nozzle are unknown, we studied the effect of the length of the constant cross section area conduit by selecting several length-to-diameter (L/d_t) ratios. The results of the simulation are plotted on a Pressure-Temperature (P-T) chart as shown in Figure 9. We could produce the same P-T chart as the Twister (Epsom, 2005) by changing the length of the conduit part by adjusting the (L/d_t) parameter. This indicates that, in addition to the nozzle outlet pressure, by adjusting the channel length, the minimum temperature of the system and shock position can be controlled.

Mesh generation

The quality of the mesh plays a significant role in the accuracy and stability of the numerical computation. The issue of grid quality is concerned with the ability of a particular discretization scheme to accurately represent the continuous governing equations on a given grid. The final accuracy and efficiency of any numerical solution are highly dependent on the particular meshing strategies and the mesh density distribution employed. The key to an efficient overall numerical solution remains in a good matching of the strengths and weaknesses of the grid generation and flow solution techniques and the maintaining of a strong and favourable interplay between these two phases of the solution procedure. Obviously, the goal of any numerical simulation should be the optimization of both the discretization scheme as well as the grid generation scheme.

Although accuracy increases with finer grids, the CPU and memory requirements to compute the solution and post-process the results also increase. Of many solutions to such deficiency -adaptive grid refinement can be used to increase and/or decrease grid density based on the evolving flow field, and thus provides the potential for more economical use of grid points and hence reduced time and resource requirements. Multigrid strategy also represents another recent powerful technique which takes a more comprehensive approach to the general problem of numerically simulating a physical phenomenon by closely coupling the grid generation and numerical solution aspects. The idea of a Multigrid algorithm, which was considered in the present numerical technique, is to accelerate the convergence of a set of fine-grid discrete equations by computing corrections to these equations on a coarser grid (Peyret, 1996), where the computation can be performed more economically. This process is applied recursively to an entire set of coarse-grid levels. Each Multigrid cycle begins on the finest grid level and cycles through the various levels up to the coarsest mesh. At this stage the computed corrections are successively interpolated back to the finest level and the cycle is repeated. The accuracy of the final discretization is solely determined by the fine-grid discretization, and the coarser levels may be viewed simply as artefacts employed to accelerate convergence.

The dependence of the accuracy on the quality of the grid can be reduced if more grid metric information is included. However, grid dependency studies (Drikakis & Tsangaris, 1993) have shown that finer meshes do not necessarily influence the accuracy of the solution in the case of the axisymmetric nozzle flow. This conclusion was also reached in our simulation when three different numbers of mesh cells, 1882, 5533 and 11 832 cells, were selected. Table 1 shows the number of iterations that led to convergence with respect to total mass error in inlet/outlet mass flow for each case. It is clearly shown that finer grid did not significantly affect the iteration steps for convergence. In fact, for coarser grids (i.e. 1882 cell), convergence occurred even faster (~1500 steps). However the position of shock changed and became stabilized after the number of iterations was doubled. The uncertainty of shock spot forces us to choose finer grids, especially since the shock wave is considered as one of the large gradient regions that requires grids to be fine enough to minimize the change in the flow variables from cell to cell. Due to the difficulties in determining the location of the shock in advance, one should strive to achieve a high-quality mesh over the entire flow domain. Hence, we considered 5533 grid cells as the computational mesh for the present study to endorse smooth variations of flow properties across shock region. Convergence occurred after a certain number of iteration steps and varied for real and ideal gas models. For example convergence for real gas non-swirl case occurred after 3182 iterations with an error in inlet/outlet mass flow of about 4.8×10^{-4} %. Table 2 shows the number of iterations that led to convergence with residual errors for continuity, energy, k and equations for non-swirl real and ideal cases.

Table 1: Number of iterations that led to convergence for simulation cases with total mass error in inlet and outlet mass flow

No. of Cells	No. of Iterations	% errors in mass flow
1 882	3 030	9.8×10^{-3}
5 533	3 182	4.8×10^{-4}
11 832	3 500	3.2×10^{-1}

Table 2: Number of iterations that led to convergence with residual errors for continuity, energy, k and equations for non-swirl real and ideal cases

Case	% Error of Inlet/Outlet Mass Balance	Number of Iteration Steps	Continuity	Energy	k	ε
Ideal	2.0×10-4	9 230	2.40×10-3	1.90×10-3	8.50×10-6	9.60×10-6
Real	4.8×10-4	3 182	5.85×10-4	5.58×10-4	5.58×10-5	6.03×10-5

De Laval Nozzles with Extended U-Shape Throat

The purpose of this section is to find alternative designs to the swirling flow of gas through the Laval nozzle configuration that was discussed earlier in search of better separation performances. As a result, the Laval nozzle with extended U-shaped section was proposed. The proposed nozzle is composed of three sections:

- The convergent subsonic section that experiences a slight divergence after the throat to accommodate for the supersonic flow of the gas.
- The U-shaped extended throat through which the centrifugal force exerted on the particles is expected to provide the intended separation between phases (gas-liquid, gas-solid, or gas-solid-liquid).
- The divergent section that provides the cross-sectional growth required for pressure recovery.

The design parameters in this study included the ratio of the inside diameter of the throat to that of the U-shaped section, the curvature radius of the U-shaped section, and the inlet and outlet pressures. System variables that were used in the analysis were position of the shock, gas velocity inside the U-shaped section, and consequently the centrifugal acceleration experienced by the gas flowing inside that section used as a measurement of the level of phase separation the device could deliver.

Simulations were based on the inviscid flow of methane as process fluid. The governing equations were those of the conservation laws (mass, momentum, and energy) along with an appropriate thermodynamic model to predict gas properties.

As no experimental results are available for evaluation purposes, a pilot test is under development that uses compressed air as process fluid. The test results will then be compared to those of the CFD simulations performed under the same conditions to evaluate the computer models.

Throat Diameter Ratio

As it was shown in the previous section, the flow always reaches sonic velocity at the throat. The slight divergence after the throat and before the U-shaped section is to allow the flow to reach supersonic velocity. Assuming pressure drop across the nozzle is large enough, the value of the Mach number at the entrance to the U-shaped section depends solely on the ratio of the diameter at this point to that of the throat. Since the purpose of the U-shaped section is to take advantage of the supersonic flow in a circular path, it is important that the shock takes place in the diverging section. It must be noted that despite the diameter of the nozzle is constant throughout the U-shaped section, a shock wave may take place inside that section due to the loss of energy in the U-shaped section.

The U-shaped section to throat diameter ratio of 1.2 was determined to be most effective in this study. This means that the constant cross-section area of the U-shaped section has a diameter of 1.2 times the diameter of the throat where the choke takes place. It is important to realize that this ratio is not the only factor that determines the position of the shock. Other parameters involved are the U-shaped section curvature radius and outlet back pressure.

U-Shaped Section Curvature Radius

The curvature radius of the U-shaped section has two converse effects on the flow. Increasing the radius of the curvature allows the gas to pass through it with faster velocity and higher Mach number without triggering a shock wave, but at the same time reduces the centrifugal acceleration (Bird, 1924). This is easily shown by the kinematic expression of tangential acceleration:

$$a_{\text{tangential}} = \frac{V^2}{r} \tag{4}$$

It may be concluded that since velocity is raised to the power of two in this equation, its value overweighs the negative value of increased radius. It should be noted though that an important feature of such supersonic separators is their relatively small size, and a rather large radius of curvature will compromise this characteristic. After vigilant study of several designs with various curvature radii, it was determined that a U-shaped section with a decreasing radius profile would be the best configuration for this design. The curvature profile shown in Figure 10 was therefore developed for the U-shaped section of the nozzle. The radius of this profile is 40 units in the beginning (lower left section) and decreases continuously until it reaches 5 units in the end (upper section).

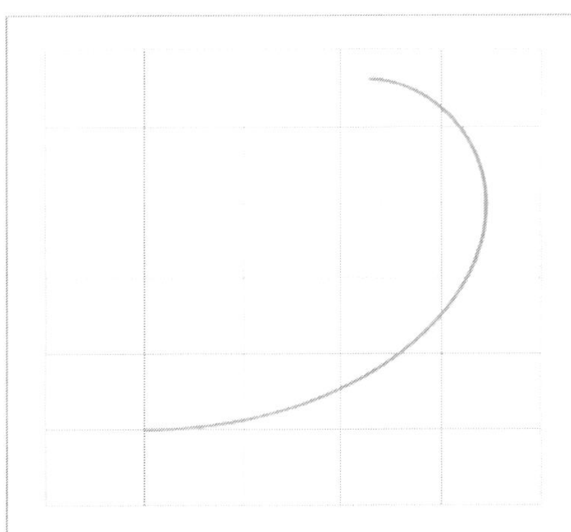

Figure 10: A typical variable curvature proposed for the U-shaped section of the Laval nozzle.

Inlet Pressure/Outlet Back Pressure

The outlet back pressure in this study was chosen to be the atmospheric pressure. The inlet pressures as well as the flow rate of the gas through the device were hence the other parameters affecting the pressure drop through the nozzle and consequently the position of the shock wave.

Particle Separation

Particle separation, as it was mentioned before, is due to the centrifugal force that is exerted on the flow as it passes through the U-shaped section with high velocity. The particles (e.g. micron size liquid droplets) are forced towards the outermost wall in the U-shaped section and may be extracted via a side channel that is installed in an appropriate position. Figure 11 shows the values of centrifugal acceleration along this section as it is experienced by the flow. The acceleration is calculated from Equation (4). The graph is generated from the CFD simulation results for the geometry shown inFigure 12 when an inlet pressure of 345 kPa (50 psia) was imposed. It can be seen in the graph that the particles experience centrifugal accelerations of up to 36,000,000 m/s² in this particular configuration. That is the equivalent of approximately 3.7 million g.

Shockwave Position

The position of the shockwave is very important for the performance of such separators. The shockwave should not take place before or through the U-shaped section as it was explained before. It is also important that the shock does not occur too far into the diverging section. This is because the farther into this section the shock takes place, it requires more differential pressure driving force across the device and therefore results in a greater pressure loss. It is crucial that the shock occurs as early as possible in the diverging section.

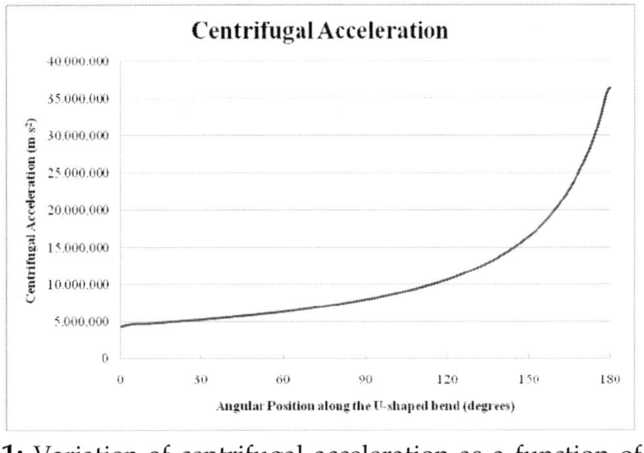

Figure 11: Variation of centrifugal acceleration as a function of position along the U-shaped section in the proposed Laval nozzle.

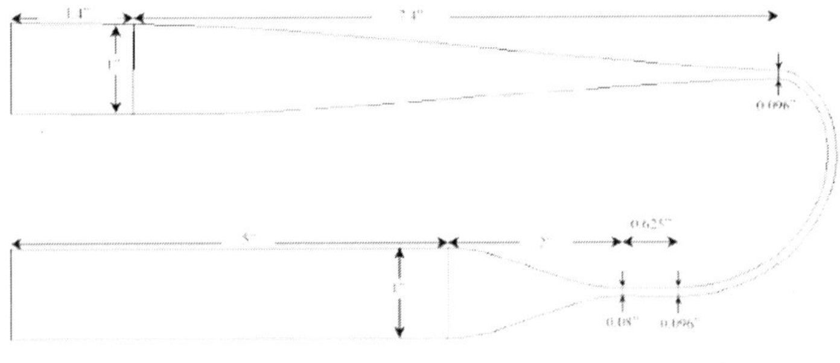

Figure 12: A typical design configuration of the proposed Laval nozzle.

This property is one of the properties that were used to adjust the design and operating conditions of the separator. The position of the shockwave can be monitored through many different methods such as pressure profiling and velocity profiling. The most effective and accurate method that became available through the CFD package's post processing capabilities was the use of Mach number contours across the device. Theses contours clearly show the position of the shockwave as well as its shape and intensity (see Figure 13).

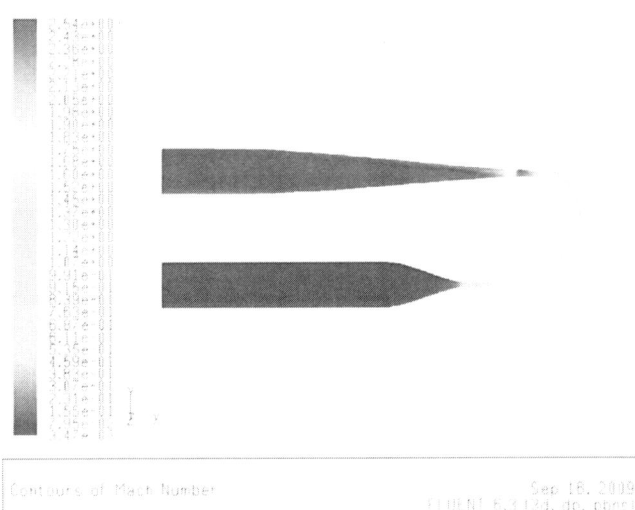

Figure 13: Mach number filled contours showing the position of the shock wave.

Separation Channel Position

One important design parameter is to determine where the side channel to separate the flow of particles from the main gas flow should be placed. This can also be easily determined using the post processing features of the CFD package. Figure 14 is a volume density contour, precisely showing where most of the higher density particles are accumulated along the U-shaped section.

Mesh Generation

As it was mentioned previously, the quality of the mesh is a key factor in the accuracy and stability of a numerical (finite volume) analysis. Various characteristic properties of the geometry itself, types of fluids involved, and the amount of available memory and processing capability are some of the parameters that contribute to the final selection of a specific mesh design. A combination of wedge shaped and tetrahedral elements was chosen to represent the geometry in Figure 12. The geometry was divided into 6 sections as it can be seen in Figure 15. All sections were meshed using wedge shaped elements except section 2 which was meshed using tetrahedral elements due to the high gradient of cross-sectional area. Section 1 is only a simple pipe and its purpose is to stabilize the flow of gas before it enters the nozzle system. The meshing of this section therefore does not require a very fine quality.

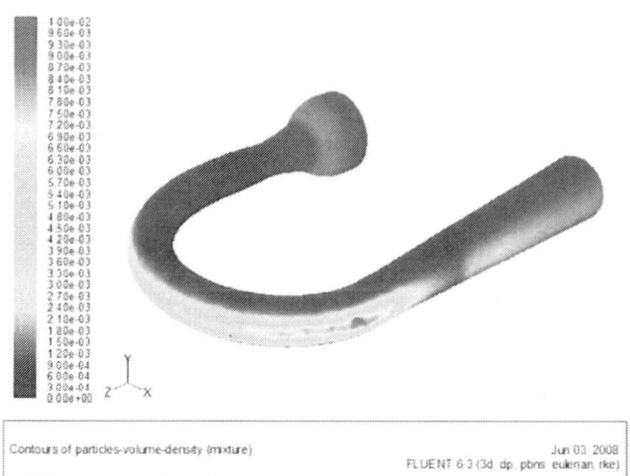

Figure 14: A typical contour of the particles volume density showing where a side channel should be positioned to separate most of the particles.

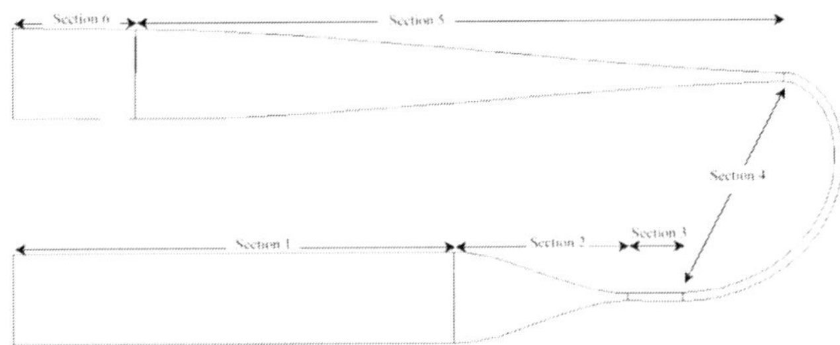

Figure 15: Geometry divisions for meshing purposes.

The mesh elements in this section are wedges with a length of 5.08×10^{-3} m (0.2″) and equilateral triangular bases with edges of 2.54×10^{-3} m (0.1″) (see Figure 16). Section 2 is the converging section of the nozzle and is meshed with a shrinking tetrahedral scheme (Figure 17 and 18). This method ensures the stability of the geometry through the numerical analysis and links the coarse mesh of section 1 to the fine mesh of section 3. Section 3 is the slight divergence after the throat that accommodates for the supersonic flow of the gas and hence requires a very fine mesh quality to ensure the accuracy of the results. The meshing scheme is again a wedge type element that grows laterally as the cross section increases. The meshes are arranged so that there are 60 elements on the cross sectional perimeter at each point. This means that the wedge bases have lateral sizes of 1.016×10^{-4} m (0.004″) in the beginning (right after the throat) and 1.27×10^{-4} m (0.005″) in the end (right before the U-shaped section). A length of 2.54×10^{-4} m (0.01″) is kept constant throughout this section. Section 4 is the U-shaped section. Wedge elements advance into this section with the same base sizes and lengths of 5.08×10^{-4} m (0.02″). The wedges grow in section 5 until they reach a base size of 2.54×10^{-3} m (0.1″) and length of 5.08×10^{-3} m (0.2″) at the end of this section to merge in with section 6 that is the equivalent of section 1 and the mesh elements remain the same throughout this section. It is important to note that since the fluids being studied are gaseous and of high velocities, the effects of boundary layers may be neglected and hence no extra care is directed towards that area.

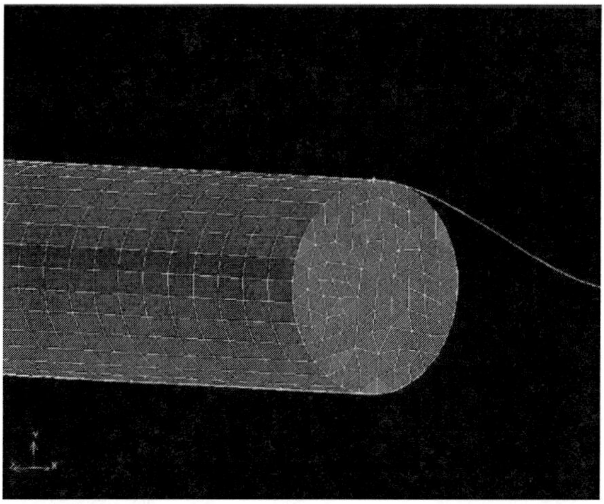

Figure 16: Wedge shaped mesh elements forming the inlet stabilizing section of the device (Section 1).

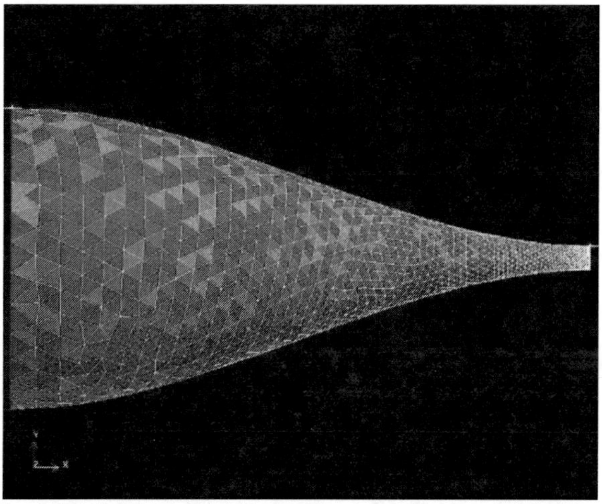

Figure 17: Tetrahedral mesh elements in the converging section of the nozzle (Section 2). The element volumes shrink as they get closer to the throat.

Experimental Apparatus

The novel nature of this study and the lack of external data inflicted the need to carry out laboratory pilot tests in order to evaluate the results of CFD simulations. The idea was to create a nozzle system similar to those that had been simulated through which high pressure gas would flow. Pressure, temperature, and flow rate measurements would be made and the results would be compared to those of the CFD simulations of the same system. The geometry chosen for this test was the geometry shown in Figure 12 and the process gas was compressed air. Two symmetrical halves of the proposed geometry was machined out of two blocks of aluminum and put together to form the desired pathways. Since it is significantly important that the flow is not disturbed by any bodily imperfection or hindrance, direct pressure and temperature measurement was not an option. Small size channels were created at several points along the pathway to enable the placement of measurement probes outside the fluid's pathway. As can be seen in Figure 19, there are 12 pressure measurement points and 4 temperature measurement points. Figure 20 shows the schematic setup of the pilot test. Measurements are currently being performed and the data is yet to be analyzed, but the preliminary results are in good agreement with the simulations.

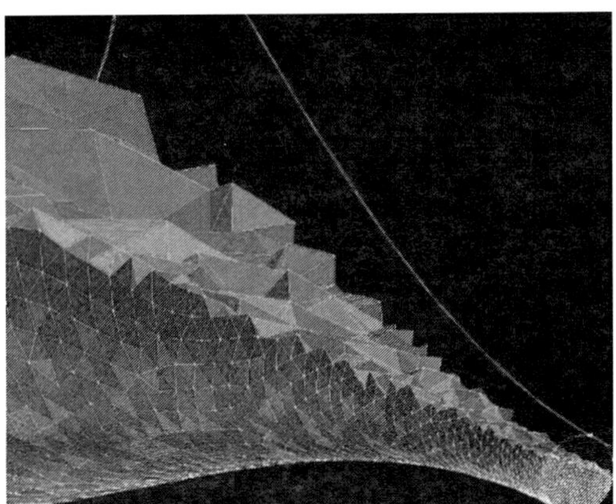

Figure 18: A planar cut view of the mesh elements in the converging section of the nozzle (Section 2).

COLD JET RELEASE FROM HIGH PRESSURE MARINE CNG TANKS

Cold jet is a result of a high pressure leak through a wall crack or valve stem or any other opening caused by an accident or failure to a high pressure device. Computational Fluid Dynamics, CFD, was used to study the phenomena and its effect on the surrounding equipment.

Problem Description

The cold jet is developed when a fluid under high pressure and quite low temperature conditions is propelled to the ambient conditions through a crack or any leak opening. The very low temperature created due to this effect can influence the material's strength of construction of the high pressure containers/vessels or pipe systems carrying the gas. Very few studies have been conducted on this area while the demand is growing in the industry for such studies.

The low temperature can make ordinary carbon steel to become very brittle and lead to instant failure of high pressure tanks or pipelines and subsequent explosion. The jet can also extend to other adjacent equipment, parts, and pieces and influence their strength and integrity due to temperature variations.

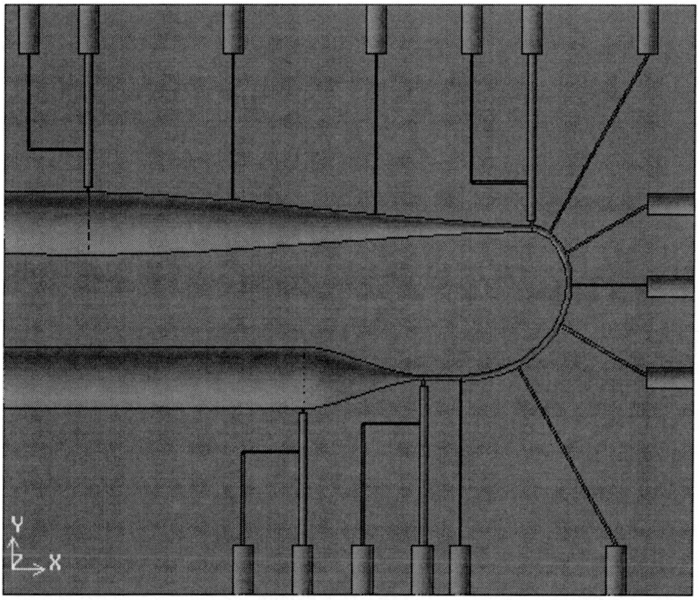

Figure 19: The bottom piece of the pilot test nozzle system. The 4 thicker channels are designed to accommodate for temperature probes. All other channels are for pressure measurements.

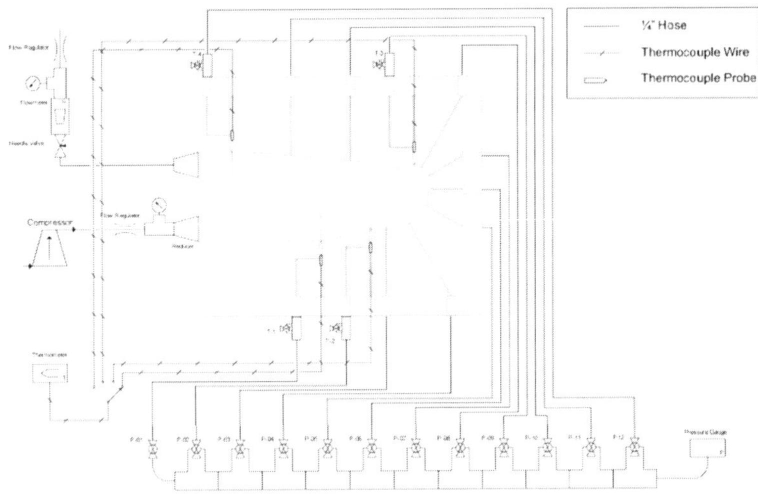

Figure 20: Process flow diagram describing the schematic setup of the test pilot set-up.

The result of such study will provide an improved insight for any failure and stress analysis when such problems are encountered.
In this study, the behaviour of flow through the crack, the temperature distribution around the crack, and the influence of the jet to the adjacent walls have been reviewed.

The computational fluid dynamics technique was used to study the behaviour of high pressure natural gas when it flows through accidental cracks. The following themes are of interest in this study and will be discussed in this section:

- Flow behaviour through the crack
- Temperature distribution around the crack
- The influence of the jet to the adjacent walls

System Configuration and Simulation

The actual shape of the crack can be very irregular however for our analyses and to simplify the simulation a convergent nozzle with the geometry shown in Figure 21 was chosen. The conditions of the natural gas (methane) in the high pressure cylinders are: pressure 122 bars and temperature -8 ^{o}C (265K), the thickness of the cylinder was 19 mm.

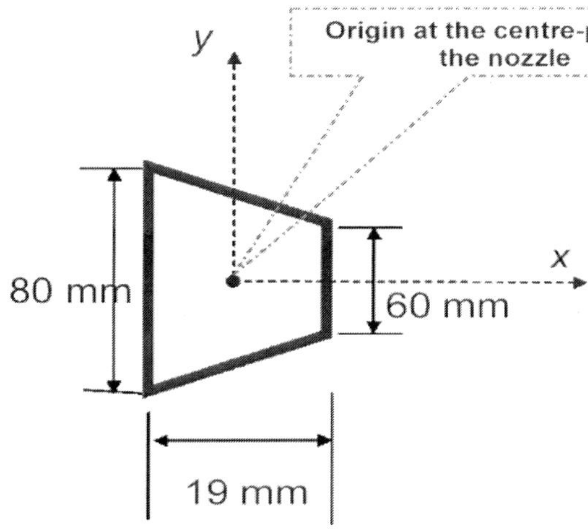

Figure 21: Crack Geometry assumed to be convergent nozzle.

Crack Simulation and Results

Figure 22 shows the Mach number variation along the crack wall. The choke flow conditions occur at the crack's exit point, which agrees well with the principles of thermodynamics. The distance is from the mid-point of the nozzle on the x-axis as seen in the Figure 21.

Wall static temperature variation shown in Figure 23 illustrates that the temperature decreases smoothly along the crack then declines very sharply near the exit, explaining the abrupt drop in pressure at crack exit.

Wall Simulation around the Crack

Temperature contours of the area around the crack can be generated by simulating the flow through the crack in a 2-D environment. Figure 24 shows the temperature contours on the outer wall surface of the tank at the crack exit. As can be seen the severe temperature differentials can pose thermal stress to the material and may cause fatigue and failure on the wall.

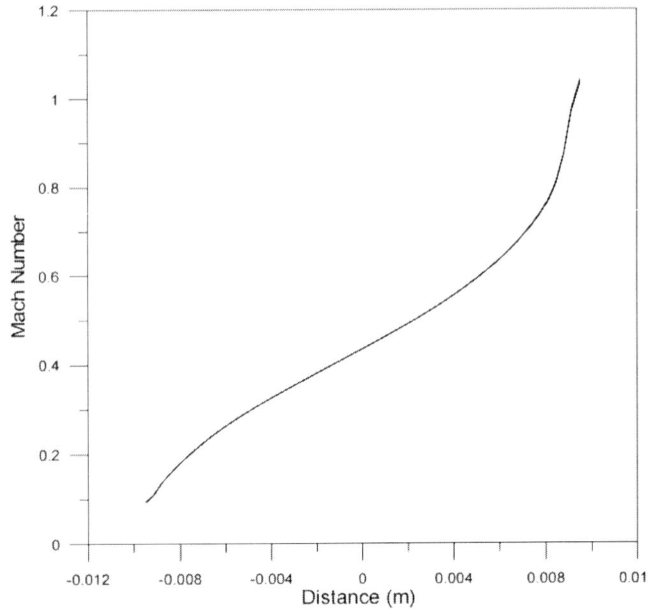

Figure 22: Flow Mach number at centerline of the Crack.

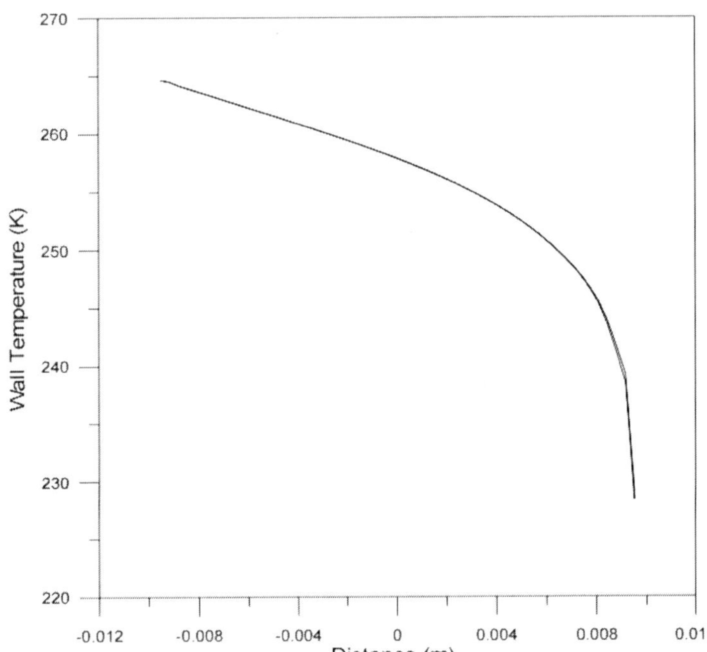

Figure 23: Temperature variation at the inner surface of the crack.

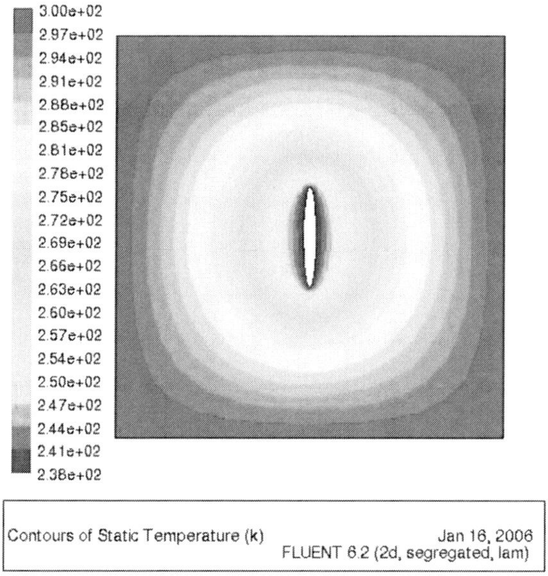

Figure 24: Temperature contours around the crack.

Jet Simulation and Results

In this section, we simulated a jet of natural gas developing outside a tank hitting a stationary wall some distant apart. Jet velocity contours are shown in Figure 25. One can recognize the impinging region and how the flow direction changes because of stagnation spot. The altered stream strikes the wall of the adjacent components (here another tank with the same height) and leaves a certain area that is exposed to impingement under less strength. Thus the temperature of this area becomes relatively higher.

It is concluded from the simulation that temperature distribution along the adjacent wall can be divided into three regions, Figures (26a, 26b):

- Impinging point vicinity
- Sharp variation region
- Mild variation region

As seen in the jet velocity profile near the adjacent wall in Figure 25, the flow slightly alters then severely changes its direction to become parallel to the wall as it moves closer. Thus there is a certain distance on the wall that stagnation or circulated flow (eddies) are in contact with the wall. Hence the temperature of this region decreases sharply from a certain value (at stagnation spot) then increases gently along the rest of distance on the wall.

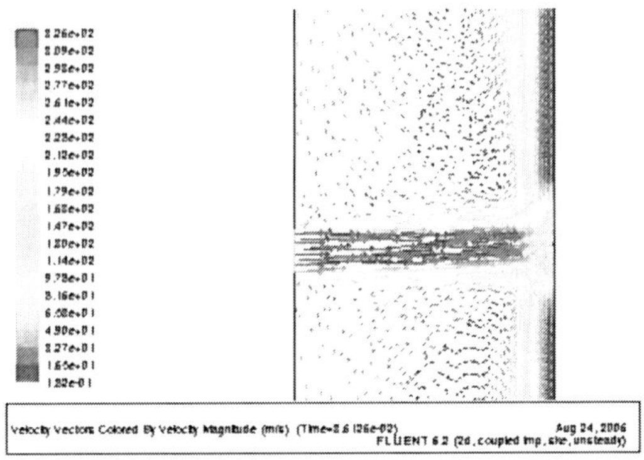

Figure 25: Jet velocity vector caused by impinging with adjacent wall.

CONCLUSIONS

Main conclusions can be summarized as follows:

Shockwave has been studied using the CFD approach. The effects of real gas properties on the flow behavior were also addressed. It is crucial to beware of the real gas fluid behavior when analyzing high pressure natural gas flow systems. The assumption of a perfect gas may give rise to significant errors. Comparison between perfect and real gas models was implemented to present the errors involved when real properties in calculating heat, work, and other thermal processes are ignored. The results of CFD simulation runs conducted for different gases showed that the ideal gas assumption would lead to serious misrepresentation of the flow field, including the position of shockwave and severe miscalculations in the design of supersonic devices.

The influence of geometry and vorticity were studied. Since the addition of a channel within the Laval nozzle is of interest in separating liquid particles, various lengths of such extended constant area regions, which connect the nozzle part with the diffuser part, have been selected. Results showed that variation of the channel length would impact the position of the shock wave and the minimum temperature of the system, which in turn influence the quantity and quality of droplets if condensation occurs. Swirls are formed when the fluid enters the nozzle semi-tangentially. Strength of vorticity and its behavior were addressed because the centrifugal force is responsible for particle separation due to the very high centrifugal acceleration induced within the nozzle throat.

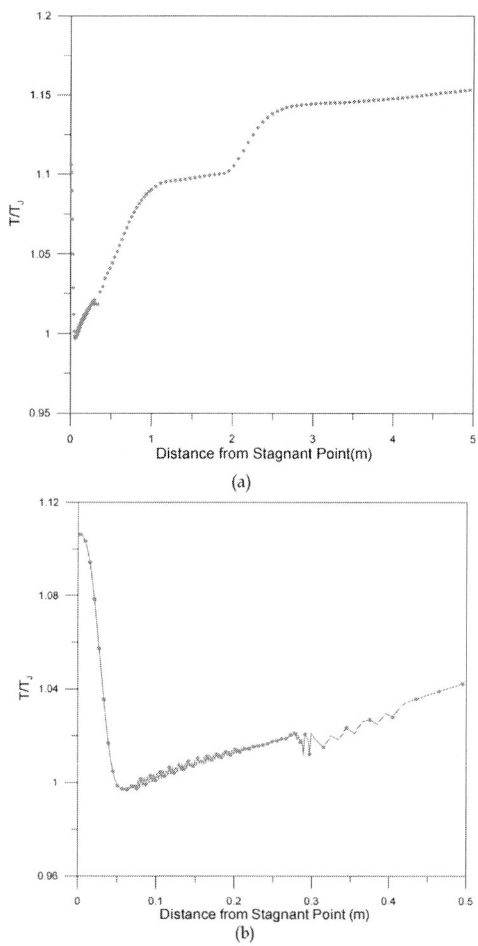

Figure 26: Wall temperature ratio as a function of distance from stagnant point: a. Temperature profile at the adjacent wall; b. Enlarged portion wall temperature profile at the stagnant region.

This acceleration is in the order of 105 or 106 times the acceleration of gravity and can induce separation of particles as small as several tenths of a macron in diameter. Furthermore, it is very important to have a sound understanding of pressure losses across the system, as these losses influence the economy of supersonic nozzles in natural gas processing. as the loss of pressure has to be compensated by costly compression facilities. In our further work, we will include the multiphase as well as multi-component fluid systems where the behavior of the natural gas in the system can be analyzed with more accuracy.

It is concluded at this point that the alternate configuration of nozzle with an extended U-shaped section can deliver better separation performances than the original swirling flow of gas in a straight nozzle. Although the centrifugal acceleration levels observed here are highly dependent on the geometry size (curvature radius), it can be speculated that very powerful centrifugal forces can be achieved in a full size separator of this type. Despite the lack of external data, the CFD simulations have proven to be significantly reliable. The practical application of this system may look far away, but the use of CFD simulations has shown great potential and encouraging results.

CFD simulation is a very efficient technique to study the flow of high pressure gas through complex geometries and sophisticated simulations such as impinging region. This study showed how CFD could be used in analyzing such phenomena with elucidation that gives it a robust position among other algorithms. The selection of the right thermodynamic model is very crucial in simulating natural gas flow under high pressure and low temperature conditions. Results show that divergent geometry of the crack changes the flow from subsonic to supersonic flow. Hence, irreversibility is clearly affecting the properties of the flow through the crack. CFD simulation for cold jet concluded that adjacent components will be affected by the jet. The temperature varies according to the impinging phenomena in both shapes, however the temperature difference between the jet and adjacent wall is relatively higher in divergent shape than convergent. Further, heat exchange due to collision of jet with the adjacent storage tanks shows significant effects on the temperature distribution.

ACKNOWLEDGEMENTS

Financial support provided by the Centre for Marine Compressed Natural Gas (CMCNG) Inc., Unconventional Gas Supply Program of the Natural Resource Canada (NRCAN), and Atlantic Canada Opportunity Agency (ACOA) to finance various process engineering projects within the CMCNG as well as the presented work in this article is appreciated.

REFERENCES

1. A. Karimi, M. Abedinzadegan Abdi, 2006 Selective Removal of Water from Supercritical Natural Gas. SPE 100442, Proceedings of the SPE Gas Technology Symposium, Calgary, Alberta, Canada, 15 18 .
2. Bai-Shi-I, 1961 An Introduction into Theory of Compressible Liquid Flow, Moscow: Izd. Inostrannoi Literatury, (Russ. Transl.)
3. C. Carvero, A. Satta, 2000 A CFD Model for Real Gas Flow. ASME paper, GT-518.
4. D. A. Kouremenos, 1986 Normal Shockwaves of Real Gases and the Generalized Isentropic Exponent. Forschung im Ingenieurwesen, 52(1), 23-31.
5. D. Drikakis, S. Tsangaris, 1993 Real Gas Effect for Compressible Nozzle Flows. Journal of Fluid Engineering. 115(117), 115-120.
6. D. Y. Peng, D. B. Robinson, 1976 A New Two-Constant Equation of State. Ind. Eng. Chem. Fundam., 15 59 64 .
7. F. Okimoto, J. M. Brouwer, 2002 Supersonic Gas Conditioning. World Oil, 223(8), 89-91.
8. G. Soave, 1972 Equilibrium Constants from a Modified Redlich-Kwong Equation of State. Chemical Engineering Science, 27(6), 1197-1203.
9. G. V. Molleson, A. L. Stasenko, 2005 An Axisymmetric Flow of a Mixture of Real Gases with a Condensing Component. High Temperature, 43(3), 422-430.
10. H. C. Man, J. Duan, T. M. Yue, 1997 Design and Characteristic Analysis of Supersonic Nozzle for High Pressure Laser Cutting. Journal of Materials Processing Technology, 63(1-3), 217-222.

11. I. P. Ginzburg, 1939 Uch. Zap. Leningr. Gos. University, 42 5
12. J. M. Brouwer, H. D. Epsom, 2003 Twister Supersonic Gas Conditioning for Unmanned Platforms and Sub sea Gas Processing Source. Offshore Europe Conference- Proceedings, Aberdeen, UK, 219 225 .
13. K. Wark, 1995 Advanced Thermodynamics for Engineers. McGraw-Hill Series, New York, USA
14. L. Crocko, 1950 One-Dimensional Consideration of the Gas Dynamics of Time-Independent Flows, in Fundamentals of Gas Dynamics, H.W. Emmons, Ed., Princeton: Princeton University Press, Princeton, USA.
15. L. Hengwei, L. Zhonggliang, F. Yongxun, G. Keyu, Y. Tingmin, 2005 Characteristic of a Supersonic Swirling Dehydration System of Natural Gas. Chinese Journal of Chemical Eng., 13(1), 9-12.
16. M. A. Saad, 1993 Compressible Fluid Flow. 2nd Edition, Prentice-Hall, Inc., New Jersey, USA.
17. O. Redlich, J. N. S. Kwong, 1949 On the Thermodynamics of Solutions. V. An Equation of State. Fugacities of Gaseous Solutions. Chem. Rev., 44 233 244 .
18. R. Arina, 2004 Numerical Simulation of Near-Critical Fluids. Applied Numerical Mathematics 51(4), 409 426 .
19. R. B. Bird, 1924 Transport Phenomena. John Wiley & Sons, New York; USA.
20. S. B. Pope, 2000 Turbulent Flows. Cambridge University Press. Cambridge, USA.
21. V. Alfyorov, L. Bagirov, L. Dmitriev, V. Feygin, S. Imayev, J. R. Lacey, 2005 Supersonic Nozzle Efficiently Separates Natural Gas Components. Oil and Gas Journal. 103(20), 53-58.

CITATION

Majid Abedinzadegan Abdi, Esam Jassim, Mohammad Haghighi and Yuri Muzychka (2010). Applications of CFD in Natural Gas Processing and Transportation, Computational Fluid Dynamics, Hyoung Woo Oh (Ed.), ISBN: 978-953-7619-59-6, InTech, DOI: 10.5772/7096.

CHAPTER 9

Incorporation of Computational Fluid Dynamics Into a Fluid Mechanics Curriculum

Desmond Adair

[1]University of Tasmania, Australia

INTRODUCTION

In this chapter, the development, implementation and evaluation of a suitable curriculum for students to use computational fluid dynamics (CFD) as part of a fluid mechanics course at intermediate undergraduate level are described. CFD is the simulation of fluids, heat transfer, etc., using modelling, that is, mathematical physical problem formulation, and numerical methods which includes, discretization methods, solvers, numerical parameters, and grid generation.Historically only Analytical Fluid Dynamics (AFD) and Experimental Fluid Dynamics (EFD) have been taught at the undergraduate level but inclusion of CFD is now possible and desirable, with the advancing improvements of computer resources.

The use of simulation can now be found in many areas of engineering education, for example for electronics laboratories (Campbell et al. 2004; Feisel& Rosa, 2002), for chemical reactions (Qian and Tinker, 2006) and for diesel engine simulation (Assanis& Heywood, 1986). Simulation has also been effective in fluid mechanics and heat transfer (Devenport&Schetz, 1998; Zheng& Keith, 2003; Rozza et al., 2009). Some work in developing an educational user-friendly CFD interface and package has already been carried out (Pieritz et al., 2004) where the general aspects and simplification of the three main processes of CFD, the pre-processor, the solver and the post-processor were carefully considered. An electronic learning system while using an existing CFD package (Hung et al., 2005), and, integration of CFD with experimentation in the area of flow control research (Ardag et al., 2009) have been reported.

CFD in Engineering Practice

Why should CFD be included in an undergraduate fluid mechanics curriculum? The simple answer is that CFD has now a major component of professional life in engineering practice and to prepare students properly they must get exposure to all aspects of their chosen profession. In the areas of analysis and design, simulation based design is commonly used instead of the traditional "build and test", as it is much more cost effective than EFD and a substantial database is provided for diagnosing the adjacent flow field. Simulations can readily be done of physical flow phenomena that are difficult to measure, for example, full scale situations, environmental effects and hazards. With the introduction of CFD into a curriculum, it is possible to educate the young engineer as to the pros and cons of using the three areas, AFD, EFD and CFD and to be in a position to know which area to use according to the problem or project confronting them. Importantly, the engineer should learn not be prejudiced against using any of the three areas. So there is an increasing need to integrate computer-assisted learning and simulation, in the guise of CFD, into undergraduate engineering courses, both as a learning tool and as initial professional training.

General Concerns about Introducing Simulation into a Curriculum

Issues of concern arise when simulation is being introduced into a curriculum. These include learning vs. research objectives, usability vs. predetermined objectives and student demographics (Stern et al., 2006). A proper balance should be sought between these competing objectives, for example, it is just as important that a student be taught the practical and systematic ways of using a CFD package in a general sense, as well as achieving a specific result. There is much evidence from previous studies that: the use of simulation enhances the curriculum (Feisel& Rosa, 2002; Rozza et al., 2009); there is increased learning efficiency and understanding (Keller et al., 2007; Kelsey, 2001; LaRoche et al., 2002); there is effectiveness of new and hands-on learning methods (Patil et al., 2009); and, it is effective to use a combination of physical and simulation laboratories (Stern et al., 2006). Importantly, user-friendly interfaces must be designed so that students do not struggle with the mechanics of performing simulations to the detriment of understanding concepts. Also, when developing a curriculum which incorporates CFD, care must be taken to include flows of current interest while including diverse learning objectives. It must be remembered that CFD has become a widely used tool in fluids engineering covering many industries including Aerospace, Chemical Processing, Automotive, Hydraulics, Marine and Oil & Gas, and, hence choices have to be made when developing a curriculum.

In engineering practice, a current pacing element is the lack of personnel capable of using CFD. Until quite recently, most engineers using CFD software in industry and research centres had completed post-graduate degrees where CFD courses were taught either formally or informally. Now, as CFD becomes more pervasive in engineering practice and engineers are expected to use CFD without post-graduate education, teaching CFD at the undergraduate has become more usual and necessary. An obvious advantage of integration of CFD software into an undergraduate lecture and laboratory course is that analytical and experimental results can be compared with CFD results. The teaching approach would be to use interactive exercises to compliment traditional fluid mechanics course, and some success has already been noted in the previously mentioned studies above.

Specific Concerns about Introducing CFD into a Curriculum

There are many issues, which if not carefully considered and implemented, can lead to teaching and learning difficulties. For example, which is best? - demonstration using CFD software, or allowing the students hands-on experience and the ability to investigate for themselves. Also, CFD could very well detract from a deeper knowledge of fluid mechanics, as, for example, boundary conditions, inlet conditions etc. are often built into the CFD package, and can be set without any real understanding. Students soon realize that they can get results, and reasonably plausible results, by mechanically following instructions, and not having much understanding of methodology and/or procedures. There is no doubt that when a student first uses CFD, a lot of new knowledge and required skills descends on them from many directions hence rendering to a steep learning curve. Without careful planning this curve can become overwhelming. Lastly, because CFD software is relatively less expensive than laboratory equipment there is a danger that it may replace laboratory experiments when this is not appropriate.

The questions in the above paragraph need to be assessed and evaluated when developing a curriculum. Here, the emphasis is on the development, implementation and evaluation of an effective curriculum for students to learn CFD, keeping in mind the issues of the previous paragraph, as part of a fluid mechanics course at intermediate undergraduate level. The objectives are to supplement and enhance the traditional course and to prepare students to use CFD effectively in engineering practice. The software chosen here is a commercial industrial software, and exposes students to the same or similar software they may use as professionals in industry. The software package provides students with a "Virtual Reality" interface, which allows for relative ease in setting up flows and the ability to visually reinforce concepts in fluid flow and heat transfer during the post-processor stage.

Outline of the Chapter

In Section 2 of this chapter, basic computational fluid dynamics elements introduced to students in two lectures at the start of their course to introduce CFD elementary theory, methodology and procedures are outlined. In Section 3 the concept of the CFD interface is explained and in Section 4 the course/laboratories, learning objectives, applications, coursework and the integration of the CFD laboratories into the existing fluid mechanics course are described. Section 5 presents evaluation design, results and discussion, in the form of three investigations, one as a controlled experiment comparing the CFD group with a controlled group, one measuring the student learned knowledge and skills regarding the CFD interface and one eliciting student views on using CFD by questionnaire. Section 6 gives conclusions drawn and possible work for the future.

BASIC COMPUTATIONAL FLUID DYNAMICS ELEMENTS

This section outlines essential elementary CFD theory, which must be introduced students before they encounter hands-on experience in the laboratory. The following are extracts fromthe initial lectures given to the CFD student group.

Overall Idea of CFD

CFD is used to replace the continuous problem domain with a discrete domain using a grid. In the continuous domain, each flow variable is defined at every point in the domain, whereas in the discrete domain, each flow variable is defined only at the grid nodes or sub-grid nodes. In a CFD solution, the relevant flow variables would only be directly solved at the grid nodes with values between obtained by interpolation. The governing partial differential equations and boundary conditions are defined in terms of the continuous variables, velocity, pressure, etc. These can be approximated in the discrete domain leading to a large set of coupled, algebraic equations in the discrete variables. Setting up this discrete system, and solving it involves a large number of repetitive calculations, hence the use of computers.

Discretization Using the Finite-Volume Method

To keep the explanation simple, consider the following 1D equation.

$$\frac{du}{dx} + u = 0; \qquad 0 \le x \le 1; \;\; u(0) = 1 \tag{1}$$

A typical discrete representation of the above equation is shown on the following grid,

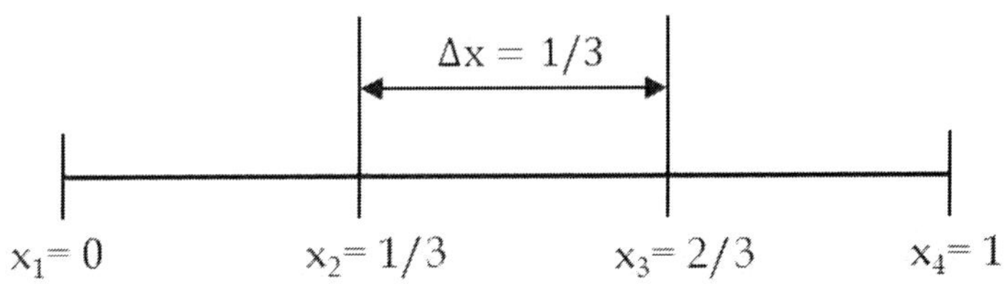

Figure 1: Discrete representation.

This grid has four equally spaced grid nodes with Δx being the spacing between successive nodes, and since the governing equation is valid at any grid node then,

$$\left(\frac{du}{dx}\right)_i + u_i = 0 \tag{2}$$

where the subscript i represents the value at grid node x_i. In order to get an expression for(du/dx)i in terms of u at the grid nodes,ui−1 is expanded in a Taylor's series,

$$u_{i-1} = u_i - \Delta x\left(\frac{du}{dx}\right)_i + O(\Delta x^2) \tag{3}$$

The error in (du/dx)idue to neglecting terms in the Taylor series is called the truncation error, and, since the truncation error is O(∅x)this discrete representation is termed first-order accurate. The following discrete equation then ensues,

$$\frac{u_i - u_{i-1}}{\Delta x} + u_i = 0 \tag{4}$$

which is an algebraic equation.

When dealing with two-dimensional geometry the grid used in the CFD laboratories here will consist of relatively simple rectangles, or a Cartesian grid. In the finite-volume method, such a rectangle is called a "cell". For 2D flows, triangular cells are often used. For the 3D flows used in the laboratories here the grid will have cuboid cells. It should be noted that it is also common to use hexahedrals, tetrahedrals or prisms. In the finite-volume approach, the integral form of the conservation equations is applied to the control volume defined by a cell to get the discrete equations for the cell. The integral form of the continuity equation for steady, incompressible flow is shown below, where the integration is over the surface S of the control volume and n is the outward normal at the surface. This really means from this equation that the net volume flow into the control volume is zero.

$$\int_S \vec{V} \cdot \vec{n} dS = 0 \tag{5}$$

Consider the rectangular cell shown,

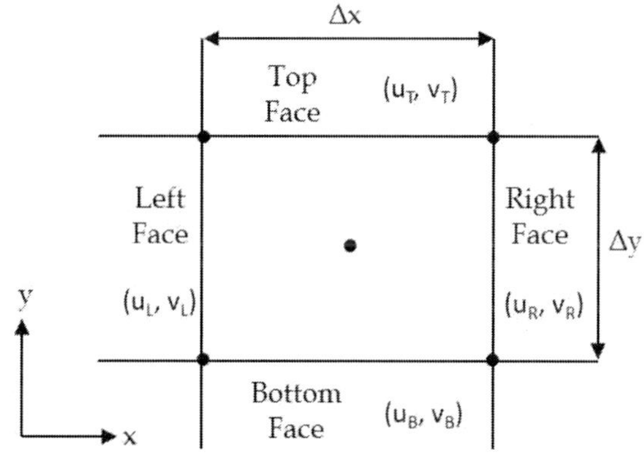

Figure 2: Cartesian control volume.

The velocity at face i is taken to be $\vec{V}_i = u_i\hat{i} + v_i\hat{j}$. Applying the mass conservation Equation (5) to the control volume defined by the cell gives,

$$-u_L\Delta y - v_B\Delta x + u_R\Delta y + v_T\Delta x = 0 \tag{6}$$

This is the discrete form of the continuity equation for the cell. It is equivalent to summing up the net mass flow into the control volume and setting it to zero. Discrete equations for the conservation of momentum and energy for the cell can also be obtained.

Rearranging Equation (4) gives,

$$-u_{i-1} + (1 + \Delta x)u_i = 0 \tag{7}$$

Applying this equation to the 1D grid (Fig. 1) shown earlier at grid points i = 2, 3, 4 gives,

$$\begin{bmatrix} -1 & 1+\Delta x & 0 & 0 \\ 0 & -1 & 1+\Delta x & 0 \\ 0 & 0 & -1 & 1+\Delta x \end{bmatrix} \begin{bmatrix} u_2 \\ u_3 \\ u_4 \end{bmatrix} = \begin{bmatrix} 0 \\ 0 \\ 0 \end{bmatrix} \tag{8}$$

The discrete equation cannot be applied at the left boundary (i = 1) since u_{i-1} is not defined. Instead, a boundary condition must be applied here.

In a general situation, the discrete equations are applied for the cells in the interior of the domain. For grid cells at or near the boundary, a combination of discrete equations and boundary equations are applied. This leads to a system of simultaneous algebraic equations. Boundary conditions are very important to have a well-defined problem and it should be remembered that a wrong boundary condition will give a totally wrong result.

In a practical CFD application, depending on the size of the grid and the number of variables to be solved for, the number of unknowns in a discrete system may run into thousands or even millions so the matrix inversion needed to solve the system of equations needs to optimized. Also due to the truncation error of Equation (3), it is clear that as the number of grid points is increased, and∅xis reduced, the error in the numerical solution would decrease and the agreement between numerical and exact solutions would be better. When the numerical solutions obtained on different grids agree to within a level of tolerance specified by the user, they are referred to as "grid converged" solutions. The concept of grid convergence applies to the finite-volume approach also when the numerical solution, if correct, becomes independent of the grid as the cell size is reduced. It is very important that the effect of grid resolution on the solution is always investigated.

Basic Equations of CFD

The Navier-Stokes and continuity equations provide the foundations for modelling fluid motion. The laws of motionthat apply to solids are valid for all matter including liquids and gases. A principal difference, however, between fluids and solids is that fluids distort without limits. Analysis of a fluid needs to take account of such distortions. The Navier-Stokes equations can be derived by considering the dynamic equilibrium of a fluid element. They state that the inertial forces acting on a fluid element are balanced by the surface and body forces. For incompressible flow, that is when the fluid density is constant, and ignoring body forces, the Navier-Stokes equations can be written as,

$$\frac{\partial u}{\partial x}+\frac{\partial v}{\partial y}+\frac{\partial w}{\partial z}=0 \tag{9a}$$

$$\rho\left(\frac{\partial u}{\partial t}+u\frac{\partial u}{\partial x}+v\frac{\partial u}{\partial y}+w\frac{\partial u}{\partial z}\right)=-\frac{\partial p}{\partial x}+\mu\left(\frac{\partial^2 u}{\partial x^2}+\frac{\partial^2 u}{\partial y^2}+\frac{\partial^2 u}{\partial z^2}\right) \tag{9b}$$

$$\rho\left(\frac{\partial v}{\partial t}+u\frac{\partial v}{\partial x}+v\frac{\partial v}{\partial y}+w\frac{\partial v}{\partial z}\right)=-\frac{\partial p}{\partial y}+\mu\left(\frac{\partial^2 v}{\partial x^2}+\frac{\partial^2 v}{\partial y^2}+\frac{\partial^2 v}{\partial z^2}\right) \tag{9c}$$

$$\rho\left(\frac{\partial w}{\partial t}+u\frac{\partial w}{\partial x}+v\frac{\partial w}{\partial y}+w\frac{\partial w}{\partial z}\right)=-\frac{\partial p}{\partial z}+\mu\left(\frac{\partial^2 w}{\partial x^2}+\frac{\partial^2 w}{\partial y^2}+\frac{\partial^2 w}{\partial z^2}\right) \tag{9d}$$

In the above equations, u, v, w are the velocity components in the x, y, z directions, ρ is the density, p is the pressure, and μ is the viscosity.

Turbulence is of fundamental interest to engineers because most flows encountered in engineering are turbulent. This happens because it is difficult to keep the flow laminar, or by intention as turbulence is essential for the engineering application. However, for turbulent flows, the variation of quantities with time is so random that its detailed variation can be of little, if any, engineering relevance so averaged quantities with time are therefore calculated using the Reynolds-averaged Navier-Stokes equations shown below.

$$\frac{\partial \bar{u}}{\partial x}+\frac{\partial \bar{v}}{\partial y}+\frac{\partial \bar{w}}{\partial z}=0 \tag{10a}$$

$$\rho\left(\frac{\partial \bar{u}}{\partial t}+\bar{u}\frac{\partial \bar{u}}{\partial x}+\bar{v}\frac{\partial \bar{u}}{\partial y}+\bar{w}\frac{\partial u}{\partial z}\right)=-\frac{\partial \bar{p}}{\partial x}+\frac{\partial}{\partial x}\left(\mu\frac{\partial \bar{u}}{\partial x}-\rho\overline{u'u'}\right)$$
$$+\frac{\partial}{\partial y}\left(\mu\frac{\partial \bar{u}}{\partial y}-\rho\overline{u'v'}\right)+\frac{\partial}{\partial z}\left(\mu\frac{\partial \bar{u}}{\partial z}-\rho\overline{u'w'}\right) \tag{10b}$$

$$\rho\left(\frac{\partial \bar{v}}{\partial t}+\bar{u}\frac{\partial \bar{v}}{\partial x}+\bar{v}\frac{\partial \bar{v}}{\partial y}+\bar{w}\frac{\partial \bar{v}}{\partial z}\right)=-\frac{\partial \bar{p}}{\partial y}+\frac{\partial}{\partial x}\left(\mu\frac{\partial \bar{v}}{\partial x}-\rho\overline{u'v'}\right)$$
$$+\frac{\partial}{\partial y}\left(\mu\frac{\partial \bar{v}}{\partial y}-\rho\overline{v'v'}\right)+\frac{\partial}{\partial z}\left(\mu\frac{\partial \bar{v}}{\partial z}-\rho\overline{v'w'}\right) \tag{10c}$$

$$\rho\left(\frac{\partial \bar{w}}{\partial t}+\bar{u}\frac{\partial \bar{w}}{\partial x}+\bar{v}\frac{\partial \bar{w}}{\partial y}+\bar{w}\frac{\partial \bar{w}}{\partial z}\right)=-\frac{\partial \bar{p}}{\partial z}+\frac{\partial}{\partial x}\left(\mu\frac{\partial \bar{w}}{\partial x}-\rho\overline{u'w'}\right)$$
$$+\frac{\partial}{\partial y}\left(\mu\frac{\partial \bar{w}}{\partial y}-\rho\overline{v'w'}\right)+\frac{\partial}{\partial z}\left(\mu\frac{\partial \bar{w}}{\partial z}-\rho\overline{w'w'}\right) \tag{10d}$$

All the instantaneous quantities were replaced by their corresponding time-averaged quantities. Also, due to the averaging process extra terms appear, for example, $-\rho\overline{u'u'}, \rho\overline{u'w'}$ appear. These terms behave like stress terms and require further equations if the system is to be solved. This will be further discussed in Section 2.6 (Turbulence Modelling) below.

Boundary Conditions

When solving the Reynolds-averaged Navier-Stokes equations and continuity equation, appropriate initial conditions and boundary conditions need to be applied. Boundary conditions are a required component of the mathematical model and direct the motion of the flow. They can be used to specify fluxes into the computational domain, with boundaries and internal surfaces represented by face zones and boundary data assigned to these zones.

Different types of boundary conditions can be applied at surfaces. When using a Dirichlet boundary condition, one prescribes the value of a variable at the boundary, e.g. u(x) = constant and when using a Neumann boundary condition, one prescribes the gradient normal to the boundary of a variable at the boundary, e.g.Cu(x)Cn=constant. It should be noted that at a given boundary, different types of boundary conditions can be used for different variables.

A wide range of boundary conditions types permit the flow to enter and exit the solution domain, for example, general (pressure inlet, pressure outlet), incompressible flow (velocity inlet, outflow), compressible flow (mass flow inlet, pressure far-field), and special (inlet vent, outlet vent, intake fan, exhaust fan). The boundary location and shape should be selected such that flow either goes in orout. This is not mandatory, but will typically result in better convergence. There should not be large gradients of variables in the normal direction to the boundary near inlets and outlets as this indicates an incorrect problemspecification. Also grid skewness near a boundary should be minimized.

Basic Numerical Solvers

In a practical problem, as mentioned above, a matrix would be extremely large, so needing a prohibitively large amount of memory to invert it directly. Therefore the matrix would be inverted using an iterative scheme instead. Iteration serves two purposes, namely, it allows for efficient matrix inversion with greatly reduced memory requirements and it is necessary to solve nonlinear equations. In steady problems, a common and effective strategy used in CFD codes is to solve the unsteady form of the governing equations and march the solution in time until the solution converges to a steady value. In this case, each time step is effectively an iteration, with the guess value at any time level being given by the solution at the previous time level. The finite-difference equation at a grid point is arranged so that the quantity to be calculated is expressed in terms of values at the neighbouring grid points, including guessed values. Then as we sweep from say left to right on the grid, successive values of the variable are updated, including any guessed values. However, since guessed values at some of the neighbouring points were used, only an approximate solution for the matrix inversion during each iteration is obtained. However as each iteration ensues across the grid, the values of the variable at each grid point converges towards the exact solution, making the error introduced due to guessing tend to zero. This iterative type of matrix inversion allows for efficient matrix inversion with greatly reduced memory requirements and it is necessary when solving nonlinear equations.

How do we judge when the solution is converged. Basically it is when the difference between the value of the variable being solved at the present iteration step and it's value solved for in the previous iteration step, referred to as the residual, is small enough. It is very common to use the summation of the residuals at each grid point normalised by the average of the variable.

Turbulence Modelling

There are two different states of flow, laminar and turbulent. Laminar flows are characterized by smoothly varying velocity fields in space and time and these flows arise when the fluid viscosity is sufficiently large to damp out any perturbations to the flow that may occur due to boundary imperfections or other irregularities. These flows occur at low-to-moderate values of the Reynolds number. Turbulent flows, on the other hand, are characterized by large, nearly random instabilities that grow until nonlinear interactions cause them to break down into finer and finer eddies that eventually are dissipated by the action of viscosity. For turbulent flow there is a deviation of the velocity from the mean value defined as,

$$u' \equiv u - \bar{u} \tag{11}$$

Due to this deviation, or more commonly called the fluctuation in velocity, and when the Navier-Stokes equations are averaged to become the Reynolds-Averaged Navier-Stokes equations, which are the equations which govern the mean velocity and pressure in the CFD package used here, extra terms are introduced called Reynolds stresses. This presents a problem in that there are more unknowns than there are equations leading to the necessity to model these extra terms to "close" the equations. There have over this last 50-60 years been many ways suggested as the best solution for closure, none of which are completely satisfactory. For the CFD laboratories here, the k-ε and the LVEL (Agonafer et al., 2008; Launder & Spalding, 1972)methods will be used.

COMPUTATIONAL FLUID DYNAMICS TEACHING LABORATORIES

Introduction

For the CFD laboratories the interface is a "three-dimensional" fully interactive environment. This interface uses, "virtual reality", is easy-to-use, and allows a student to simulate a flow from beginning to end without resorting to specialized codes. The code is also used widely in the professional engineering, so giving students useful skills which contribute to their preparation for the workplace. The virtual reality environment is designed as a general purpose CFD interface consisting of the VR-Editor (pre-processor), the VR-Viewer (post-processor) and the solver module which performs the flow simulation calculations. The VR-Editor allows the student to set the size of the computational domain, define the position, size and properties of objects to be introduced into the domain, specify the material(s) which occupies the domain, specify the inlet and outlet boundary conditions, specify the initial conditions, select a turbulence model when necessary, specify the position and fineness of the computational grid, and specify other parameters influencing the speed of convergence of the solution procedure. On setting up a particular flow, the characteristics of the flow are calculated using the solver module. The progress of the calculations is clearly monitored until convergence is reached or the iteration limit is reached. In the VR-Viewer, the results of a flow simulation are displayed graphically. The post-processing capabilities of the VR-Viewer used here are, vector plots, contour plots, iso-surfaces, streamlines and x-y plots.

Interface Design Specifications

The CFD laboratory is designed so that practical procedures are user-friendly and easy to implement, and also to show students that

CFD methodology needs to be systematic and rigorous. The complete process, at this level of CFD can, if so desired, be completely automated with the students going through a step-by-step process seamlessly from the set-up of the problem, through the solving to the display of the results. However it is very important that the laboratory also mirrors what is found in engineering practice, where a systematic approach is found. Careful consideration must be given to the areas listed in Table 1 below.

Table 1: Areas for systematic consideration

Geometry	Solid and other fluid boundaries
Physics	Incompressible/compressible fluid, which quantities to be solved for, closure of the equations, initial and boundary conditions
Grid	The choice here is Cartesian meshing or orthogonal meshing. The Cartesian mesh can be automatically generated or built manually.
Numerics	Convergence monitoring, selection of numerical scheme.
Post-processing	Flow visualization, analysis, verification, validation using published experimental or empirical data.

To contribute to the student's self-learning, a hierarchical system of predefined active options within the virtual reality environment are introduced for later simulations.

Interface Design Features

The design of the CFD interface (shown on Fig. 3) was chosen and designed to have features which systematically informs, is vocationally sound and is easy to use.

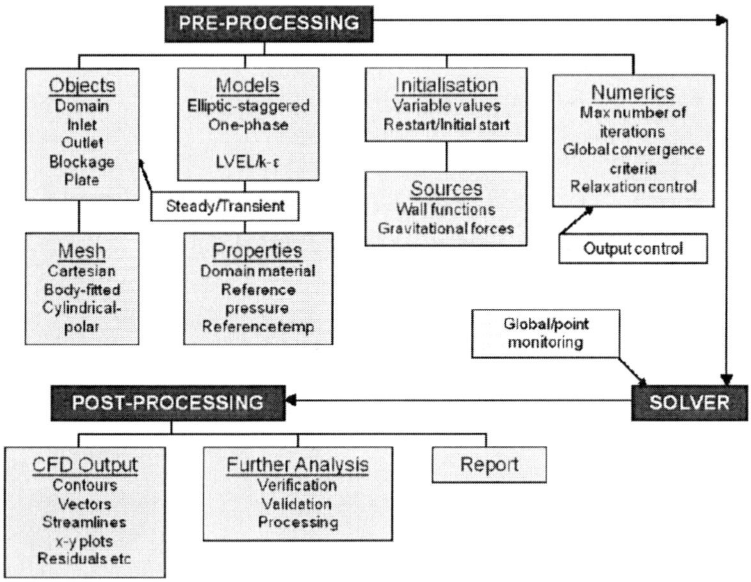

Figure 3: Summary of CFD interface.

Each simulation process follows exactly how modelling is carried out in engineering practice, with the students setting up, solving and analysing the particular problem step-by-step.

The students interact with the software using mouse and keyboard input. There was no requirement for advanced computer language skills enabling the students to concentrate on the methodology and procedures of CFD. An important feature of the CFD interface is that it is stand-alone. By this is meant that grid generation, solving and post-processing are all combined in the virtual reality environment.

Also important is that the results obtained by students could be easily moved to Microsoft Office during the reporting stage. As the software package is built on the Windows OS using PCs with relatively low computer power, it was important that the CFD interface was built on fast and accurate solvers, as is found for this software. Because students are given a limited time in the laboratory, and also in order to keep their level of interest heightened, results should come back reasonably quickly. The post-processor was designed to plot contours, vectors, streamlines and, when needed, to make animations. Students had access to colour printers to produce figures for their reports.

INTEGRATION OF CFD LABORATORY INTO FLUID MECHANICS CURRICULUM

Existing Fluid Mechanics Undergraduate Course

The CFD laboratory was integrated into a fourth semester course, for students of mechanical engineering. Traditionally the course used two lectures, each 3 hours in length, per week for theoretical fluid mechanics with four experimental laboratories, from Week 3 to Week 10, giving hands-on experience and demonstration of fundamental principles over the twelve week semester. The laboratories were given 3 hours per week to complete. With the introduction of CFD, the course was restructured to consist of two lectures, each three hours per week for theoretical mechanics, and, two CFD laboratories and two experimental laboratories. Again these laboratories were from Week 3 to Week 10, and the CFD laboratories alternated with the experimental laboratories. Two textbooks were also added to the required reading list (Ferziger&Peric, 1996; Tu et al., 2007).

The main learning outcomes are to understand the equations that govern fluid flow (conservation of mass, momentum and energy) and be able to apply them to a range of practical problems, including:

- predicting drag forces on bluff, streamline bodies and flat plates;
- analysing the flow in pipe systems;
- analysing performance of radial flow pumps and turbines; and,
- matching pumps and turbines for particular applications.

The unit also aims to develop skills in working effectively with others through the laboratory component of the unit. One or two seminars were help early in the semester, to discuss expectations regarding laboratory practice and reporting. The CFD student group was initiated to CFD as discussed next.

Necessary Initiation of Students to CFD

With the introduction of CFD into the course two extra lectures, each 2 hours in length, entitled "An Introduction to Computational Fluid Dynamics (CFD)" were presented to prepare students to learn CFD methodology and procedures.

Complementary Nature of CFD and Experimental Fluid Mechanics

During the first lecture, the students are introduced to the idea that theoretical fluid mechanics, experimental fluid mechanics and computational fluid dynamics are complementary in modern engineering practice. As with the experimental laboratories, students are then introduced to CFD general methodology and procedures. The students learn as to when and why CFD is used, and the breakdown of CFD into three processes namely the pre-processor, the solver and the post-processor. This is followed by the idea of a finite domain, subdividing the domain into control volumes (Cartesian grid), selecting the quantities to be calculated (and hence which conservation equations to be used), geometrical aspects, fluid and solid properties, sources within the domain (e.g. gravitational force), some of the numerical details (including initialization, the iterative process and how to achieve convergence), the use of different boundary conditions (solid wall, inlet, outlet), and finally how to close the conservation equations using turbulence modelling (k-ε, LVEL) (Agonafer et al., 2008;Launder & Spalding, 1972). Lastly, and importantly, the students are taught to be critical of their results and how to examine if what they are getting is what they might have expected.

CFD Methodology and Procedures

In the second lecture, the important part is a demonstration with full facilities for students to have 'hands-on' experience as the demonstration proceeds. Several simple three-dimensional flows are used as exemplars to give an overall view of the CFD process. The virtual reality environment, (Fig. 4), which facilitates the VR-Editor (pre-processor), the VR-Viewer (post-processor) and the solver module, which performs the simulation calculations, is introduced.

It is demonstrated how the VR-Editor is used to set the size of the computational domain, defining the position, size and properties of objects introduced into the domain, specifying the material which occupies the domain, specifying inlet and outlet boundary conditions, specifying initial conditions, selecting a turbulence model (if appropriate), specifying the fineness of the computational grid and setting parameters which influence the speed of convergence of the solution procedure. Viewing movement controls are also introduced for zooming, rotation, and, vertical and horizontal translations. It is then shown how to create a new simulation and to set data for geometry, automatic meshing, the quantities to be solved and the turbulence model. Objects are introduced as required together with any inlets and outlets. Finally the solver parameters are specified for the total number of iterations, the position of a 'probe', partially needed for monitoring convergence as the solver proceeds, is set, and the students are encouraged to have a final look at the automatic mesh and positions of objects before starting to solve.

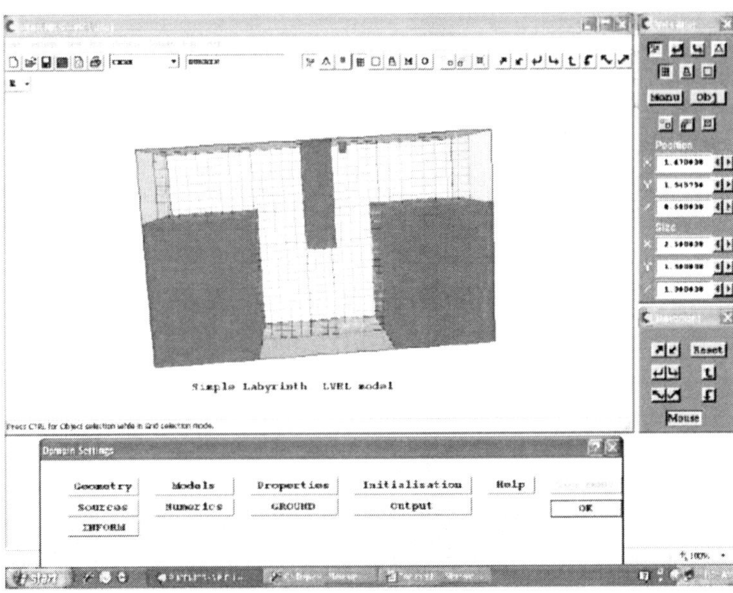

Figure 4: General view of VR-Editor (pre-processor).

As a converged solution is approached, the variables at the monitoring point become constant while with each successive sweep through the domain, the values of the errors are shown to decrease steadily.

On completion of the solver, the results of the flow-simulation can be viewed using the post-processor called the VR-Editor. This can display vector plots, contour plots, iso-surfaces, streamlines and x-y plots as demonstrated on Fig. 5. For ease of use the VR-Viewer is close to the design of the VR-Editor but with clearly marked toggle-buttons for vector display, etc.

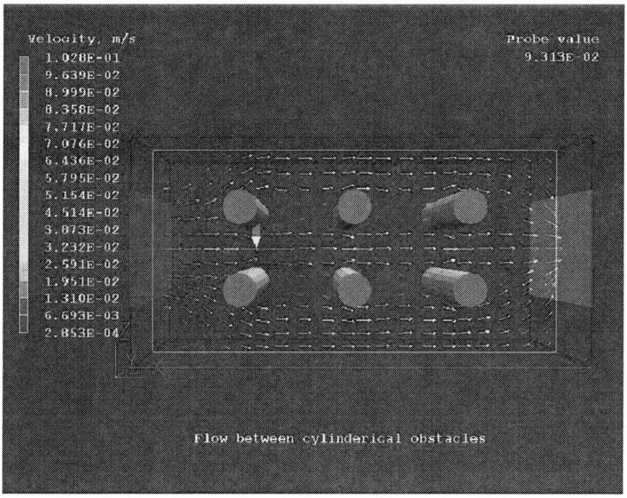

Figure 5: Typical vector-plot output of the post-processor.

The students are encouraged to display the quantities calculated in various ways and also to experiment with each of the control buttons for zooming, rotating and obtaining meaningful views of selected planes. The students are shown how to print their results and save both input and output files. A second example is then demonstrated where instead of using the automated mesh option, students are taught to think why and where fine and coarse meshing is appropriate and how to implement it. Also included is how to produce x-y plots, an important part of assignment work, and reporting system in professional life.

This lecture concludes with a broader look at CFD, including a more in-depth look at numerical methods in CFD, turbulence modelling for CFD and grid-generation alternatives to the Cartesian grid.

Replacing Laboratory Experiments with CFD Simulation
From the original four experimental laboratories, "Predicting drag forces on bluff, streamline bodies and flat plates" and "Analysing the flow in pipe systems" were chosen to become CFD laboratories while "Analysing performance of radial flow pumps and turbines" and "Matching pumps and turbines for particular applications" remained as experimental. The four laboratories were conducted sequential from the beginning to the end of the semester. The students were expected to be self-guided to encourage self-learning for both the CFD and experimental laboratories, although a tutor and technician were in attendance. Detailed step-by-step notes were provided for all four laboratories.

To overcome some of the issues mentioned in the introduction concerning integration of CFD into the fluid mechanics course, it was decided to allow students, as much as possible, to have hands-on experience and investigations and assignments by themselves as opposed to demonstration. When setting boundary conditions, especially near-wall conditions, students had to explain such topics as the need for grid refinement, so getting them to think rather than just mechanically do. Also the students were encouraged to gradually develop a 'feeling' for when pressure or mean velocity gradients were high within the flow and hence generate higher mesh densities, so again getting away from a mechanical approach. The steep learning curve met during the early days of the course was partly alleviated by getting the students to stick strictly to the procedures of Fig. 3.

TEACHING AND LEARNING EVALUATION

The evaluation process was subdivided into three investigations, one in the form of a controlled experiment comparing the knowledge of the group with CFD in their course with those of a controlled group using only the conventional experimental laboratories, the second measuring student knowledge and skill outcomes for the CFD interface, and, the third in the form of an online questionnaire eliciting the views of students on using CFD.

Controlled Experiment

To investigate the effectiveness of introducing CFD laboratories into the fluid mechanics course, a controlled experiment applying a pre-test-post-test control group design was conducted (Pfahl et al., 2004). The students had to undertake two tests, one before the respective course (pre-test) and one after the respective course (post-test) with the introduction of CFD laboratories then being evaluated by comparing within-student post-test to pre-test scores, and by comparing the scores between students in the CFD group (A), i.e. those who were taught using the course containing CFD laboratories, to those students in the control group (B), i.e. taught using the conventional method of experimental laboratories only. The various possibilities of the methods of teaching the students are summarized using Fig. 6.

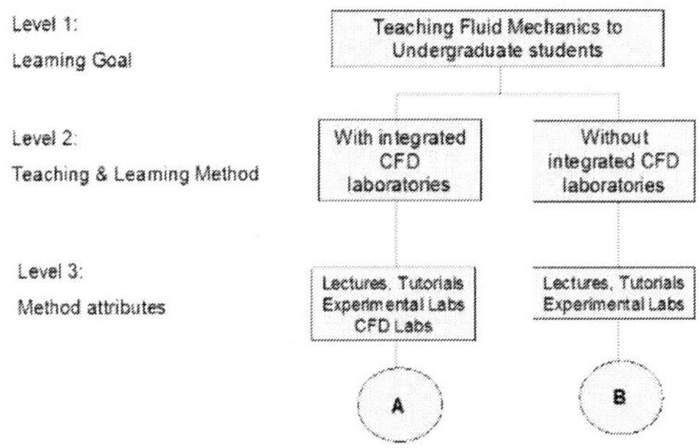

Figure 6: Course arrangements.

To measure the performance of the two groups, four constructs were used with each construct represented by one dependent variable. Each dependent variable has the hypothesis,

1. There is a positive learning effect in both groups (A: CFD group, B: controlgroup). This means post-test scores are significantly higher than pre-test scoresfor each dependent variable.

2. The learning is more effective for group A than for group B, either with regard tothe performance improvement between pre-test and post-test (the relativelearning effect), or with regard to post-test performance (absolute learningeffect). The absolute learning effect is of interest because it may indicate anupper bound of the possible correct answers depending on the method ofteaching.

The design starts with random assignment of students to the CFD group (A) and control group (B) with the members of both groups completing a pre-test and post-test. The pre-test measured the performance of the two groups before the courses and the post-test measured the performance of the two groups after the courses. The students did not know that the post-test and pre-test questions were identical and neither were they allowed to retain the pre-test questions with the correct answers only given to the students after the experiment.

The students were in the fourth semester of an engineering course with the number of students in group A, $N_A = 46$, and in group B, $N_B = 35$. The personal characteristics of the students are summarized in Table 2.

Table 2: Personal characteristics

Characteristics	
Average age	22.3 years
Percentage female	21%
Preferred learning style(s)	
Reading with exercise	16%
Lecture	11%
Tutorial	30%
Laboratory	18%
Working in groups (with peers)	25%
Opinion of most effective learning style(s)	
Reading with exercise	9%
Lecture	10%
Tutorial	32%
Laboratory	18%
Working in groups	31%

The initial testing was conducted after a short introduction as to the purpose of the experiment and general organizational issues. The pre-test was then carried out with the data for the dependent variables collected. Following the pre-test, the students were placed in either the control group or the experimental group and all students participated in both the pre-test and post-test. After completing their courses, both groups of students performed the post-test using the same questions as during the pre-test, thus providing data on the dependent variables for the second time. In addition the students were asked to answer questions about subjective perceptions.

Table 3: Experimental variables

Dependent variables
J.1 Interest in Fluid Mechanics ('Interest')
J.2 General knowledge of Fluid Mechanics('Understand general')
J.3 Understanding of 'simple' Fluid Mechanics ('Understand simple')
J.4 Understanding of 'difficult' Fluid Mechanics ('Understanding difficult')
Subjective perceptions
S.1 Available time budget versus time need ('Time pressure')
S.2 Course evaluation (useful, engaging, easy, clear)

Data for two types of variables were collected, the dependent variables (J.1, …, J.4) and the subjective perception variables (S.1, S.2). These variables are listed in Table 3. The dependent variables are constructs used to capture aspects of learning provided by the courses and each was measured using 5 questions. Selected examples of questions used as shown in Table 4.

The results for the dependent variable J.1 were found by applying a five-point Likert-type scale (Likert, 1932) with each answer mapped to the value range R= [0, 1].

The values for variables J.2 – J.4 are average scores derived from five questions for each. Missing answers were marked as incorrect. The data for the subjective perception variables was collected after the post-test. The values for variable S.1 are normalized averages reflecting the time needed for understanding and doing the tasks associated with Weeks 2 – 12.

The descriptive statistics for the experiment are summarized in Table 5. The columns 'Pre-test scores' and 'Post-test scores' show the calculated values for mean($\bar{x}$), median (m) and standarddeviation (σ)of the raw data collected, and the column 'Difference scores' shows the difference between the post-test and pre-test scores.

Table 4: Example questions (pre-test, post-test, subjective perceptions)

J.1 example question I consider it very important for mechanical engineering students to know as much as possible about fluid mechanics. (1 = fully agree / 5 = fully disagree) Circle number below. Agree 1 2 3 4 5 Disagree
J.2 example question What are the four main forces acting on an aircraft when flying straight and level?
J.3 example question What is flow separation? What causes it? What is the effect of flow separation on the drag coefficient?
J.4 example question Fluid flows out of a large tank into a straight section of pipe with a diameter d. A boundary layer of thickness δ grows along the pipe wall. Transition occurs at $x = 0$ due to a sharp edge at the inlet. The turbulent boundary layer development may be approximated by the flat-plate relation, $\delta/x = 0.391\ Re^{-0.2}$. Estimate the distance required for the boundary layer to completely fill the pipe for a Reynolds number of 2×10^5, neglecting changes in core velocity U with x.
S.1 example question I did not have enough time to: - complete the tutorials - complete the laboratory sessions - write-up the laboratory reports - complete the post-test
S.2 example question I consider the explanations/information provided for Laboratory Sessions 1 2 3 4 5 Useful Useless Boring Engaging Difficult Easy Clear Confusing

Table 5: Scores of dependent variables

	Pre-test scores				Post-test scores				Difference scores			
	J.1	J.2	J.3	J.4	J.1	J.2	J.3	J.4	J.1	J.2	J.3	J.4
Group A (x̣) (m) (ɔ)	0.67 0.71 0.11	0.61 0.60 0.34	0.27 0.24 0.29	0.23 0.24 0.25	0.69 0.72 0.09	0.83 0.78 0.22	0.58 0.49 0.21	0.35 0.31 0.27	0.02 0.01 0.01	0.22 0.18 0.29	0.31 0.25 0.25	0.12 0.07 0.26
Group B (x̣) (m) (ɔ)	0.76 0.75 0.09	0.52 0.54 0.18	0.34 0.31 0.32	0.21 0.18 0.31	0.77 0.78 0.13	0.57 0.56 0.21	0.55 0.53 0.13	0.37 0.39 0.22	0.01 0.03 0.11	0.05 0.02 0.19	0.21 0.22 0.24	0.16 0.21 0.27

Table 6 shows the calculated values for mean, median and standard deviation of the raw data collected on subjective perceptions.

Table 6: Scores of subjective perceptions

	S.1	S.2
Group A (x̣) (m) (ɔ)	0.54 0.52 0.19	0.49 0.49 0.13
Group B (x̣) (m) (ɔ)	0.46 0.41 0.21	0.48 0.44 0.14

The students in control group (B) expressed less need of additional time than those of the CFD group (A), while students of both groups were fairly equal in their perception of their respective course usefulness, engagement, difficulty and clarity.

Standard significance testing was used to investigate the effect of the treatments on the dependent variables J.1 to J.4. The null hypotheses were,

- $H_{0,1}$: There is no difference between pre-test scores and post-test scores within experimental group (A) and control group (B).
- $H_{0,2a}$: There is no difference in relative learning effectiveness between CFD group (A) and control group (B).
- $H_{0,2b}$: There is no difference in absolute learning effectiveness between CFD group (A) and control group (B).

For $H_{0,1}$ and focusing on the CFD group (A), Table 7 shows the results using a one-tailed t-test for dependent samples. Column one specifies the variable, column two represents the Cohen effect size, d, (Cohen, 1988; Ray &Shadish, 1996), column three the degrees of freedom, column four the t-value of the study, column five the critical value for the significance value $\alpha = 0.10$ and column six lists the associated p-value.

Table 7: Results for 'post-test' versus 'pre-test' for group A

Variable	d	df	t-Value	Crit.$t_{0.90}$	p-Value
J.1	0.2	45	1.36	1.301	0.09
J.2	0.77	45	5.22	1.301	0
J.3	1.22	45	8.27	1.301	0
J.4	0.46	45	3.11	1.301	0.001

It can be seen from Table 7 that all dependent variables achieve a statistically and practically significant result.

Table 8 shows the results of testing hypothesis $H_{0,1}$ for the control group (B) using a one-tailed t-test for dependent samples. The structure of the table is the same as that of Table 7.

Table 8: Results for 'post-test' versus 'pre-test' for group B

Variable	d	df	t-Value	Crit.$t_{0.90}$	p-Value
J.1	0.1	34	0.53	1.307	0.299
J.2	0.3	34	1.54	1.307	0.066
J.3	0.9	34	5.08	1.307	0
J.4	0.6	34	3.55	1.307	0

It can be seen from Table 8 that the control group (B) achieved statistically and practically significant results for dependent variables J.2 – J.4. For J.1 no significant results can be found.

For $H_{0,2a}$ which states that the difference between post-test and pre-test scores of group A is not significantly larger than the one for group B. Table 9 shows for each dependent variable separately the results of testing hypothesis $H_{0,2a}$ using a one-tailed t-test for independent samples.

Table 9: Results for 'performance improvement' (Group A versus Group B)

Variable	d	df	t-Value	Crit.$t_{0.90}$	p-Value
J.1	0.13	79	1.16	1.292	0.125
J.2	0.69	79	6.13	1.292	0
J.3	0.41	79	3.64	1.292	0
J.4	-0.15	79	-1.33	1.292	0.906

It can be seen that the hypothesis $H_{0,2a}$ can be rejected only for the variables J.2 and J.3. The value for J.4 indicates that the relative learning effect is superior when the students were exposed to experimental laboratories only.

Table 10 shows for each dependent variable separately the results of testing $H_{0,2b}$ using a one-tailed t-test for independent samples.

Table 10: Results for 'post-test improvement' (Group A versus Group B)

Variable	d	df	t-Value	Crit.$t_{0.90}$	p-Value
J.1	-0.72	79	-6.43	1.292	1
J.2	1.21	79	10.82	1.292	0
J.3	0.17	79	1.52	1.292	0.066
J.4	-0.08	79	-0.715	1.292	0.762

Again the two variables which show statistically significant results are J.2 and J.3 and hence $H_{0,2b}$ can be rejected for these variables. The variables J.1 and J.4 indicated that more interest is found in the totally experimental course and these students also did better in the more difficult aspects of the course.

Student Knowledge and Skill Outcomes for the CFD Interface

An objective measure of student knowledge and skill outcomes for the CFD interface as applied to the fluid mechanics curriculum was devised. Some of the questions used in the test are shown in Table 11, with the questions directed only at students of CFD group (A). This test was again run on a pre/post CFD studies basis, i.e. during the first week of the course students completed the pre-test and later in the semester, and after completing the CFD studies, the students completed the post-test. The most intuitive test of students' knowledge and skill outcomes is whether the post-test scores were significantly higher than those of the pre-test scores. Table 12 contains the results for the mean and variance, the number of students (N) taking the test is also shown and the test contained 20 questions.

As can be seen from Table 12, students correctly answer about 36% prior to being instructed in CFD and about 80% average correct for the post-test.

This represents a considerable improvement and is statistically highly significant, i.e.

$$\bar{x}_{post} - \bar{x}_{pre} = 8.52; \quad t(1,33) = 15.3; \quad p < 0.0001$$

It can be seen that the effect is substantial between pre- and post-tests and therefore represents significant improvement in outcomes of the students' knowledge and skills of CFD knowledge and skills. The students, after a relatively brief exposure to and with limited practice of CFD have shown considerable growth in their understanding of CFD concepts, principles and applied problems.

Table 11: Examples of test questions

Question No.	Question
1	For flow over a cylinder, what is the cause of the different results found for CFD and in the experimental laboratory? a. The difference is caused by the experimental laboratory uncertainties. b. The difference is caused by the errors from numericaland experimental laboratory uncertainties. c. The difference is caused by the errors from numerical methods. d. The difference is caused by the errors from numerical, modelling and experimental laboratory uncertainties.
2	What is a typical CFD process? a. Geometry → Mesh → Properties → Models → Initiation → Verification → Sources → Numerics → Solver → Post-processing b. Mesh → Geometry → Properties → Models → Initiation → Sources →Numerics → Solver → Post-processing → Verification c. Geometry → Mesh → Properties → Models → Initiation → Sources →Numerics → Solver → Post-processing → Verification d. Geometry → Models → Mesh → Properties → Initiation → Sources →Numerics → Solver → Post-processing → Verification
3	What is the criterion for the validation of a CFD simulation by a laboratory experiment? a. If the difference between the CFD and experimental data is less than the convergence limit. b. If the difference between the CFD and experimental data is less than the CFD data uncertainties. c. If the difference between the CFD and experimental data is less than the experimental data uncertainties. d. If the difference between the CFD and experimental data is less than the combination of the experimental and CFD data uncertainties.

Table 12: Mean number of correct answers (out of 20)

Pre-test		Post-test		N
x̣pre	ʒ2pre	x̣post	ʒ2post	35
7.35	5.28	15.87	4.94	

The success of this introduction could not be assessed using a student comparison performance in CFD laboratories across the years as these data were not available or not in a form that would make for meaningful comparison. This is not necessarily a weakness of the study as it has been suggested (Lucas, 1997) that the measurements of differences in student assessment over time has limited value given the changing nature of student cohorts from one year to another. On the basis of these recommendations a questionnaire was developed to elicit perceptions of the introduction of CFD from students involved in the class.

Online Questionnaire

An anonymous online survey was conducted after students obtained their grades for the laboratory reports to aid formative evaluation of the introduction of CFD. Only students who had completed the course with CFD were surveyed. A questionnaire using 11 statements as listed in Table 13 was designed for this survey. Students were requested to respond to each item in the questionnaire using a five-point scale: strongly agree, agree, neutral, disagree and strongly disagree plus a column for no opinion. An opportunity was also provided for students to comment on their experience at the end of the questionnaire to collect qualitative feedback on their experience so far with CFD.

Table 13: A list of questions/statements used in the survey for students' feedback

No.	Question/Statement
1	I found the software easy to use.
2	I have used CFD modelling before.
3	This CFD tool enhances my understanding of the theory course.
4	This CFD tool is a useful addition to the fluid mechanics laboratories.
5	The 'hands-on' aspects of the CFD tool has taught me extra skills.
6	The 'hands-on' aspects of the CFD tool has given me deeper knowledge of fluid mechanics.
7	On using CFD I have learned things that could not be taught through the theory or experimental courses.
8	I now have a knowledge of CFD procedures and methodology.
9	I feel I could now continue to model basic flows.
10	On completion of this course I have run at least one flow simulation with the software provided.
11	I would recommend the CFD laboratory to others.

Generally, student feedback surveys have a very low response rate (Gamliel&Davidovitz, 2005; Nulty, 2008). However the response rate here was high (>80%) with 5 responses received per question and overall, the results from the survey were positive. The responses to the survey are shown on Fig. 7 and indicate that students felt that they benefited from their exposure to CFD. In additional comments most of the students expressed the view that the amount of material introduced was correct, although some felt that the exercises took a long time to complete correctly. Students were particularly appreciative that they could easily visualize flow using contour and vector plots and generally agreed that the combination of theory, experimental and CFD led to better understanding of fluid mechanics. Students also showed enthusiasm for learning more about CFD.

In addition to the questionnaire of Table 13, the students were asked "Would you recommend that CFD remains in the fluid mechanics course in the future?" To this 82% said yes so showing that they thought CFD as having a positive impact on their studies.

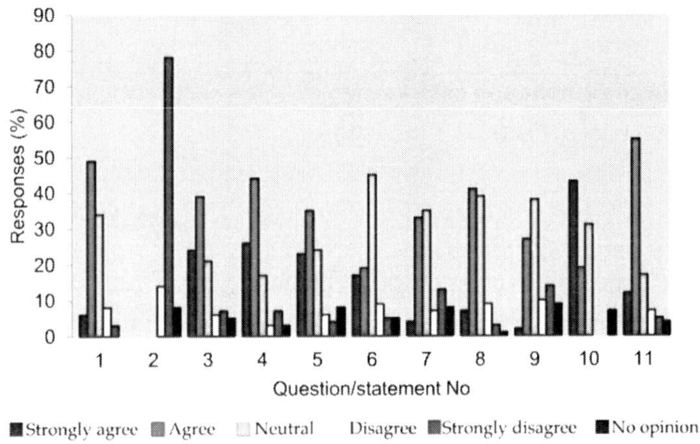

Figure 7: Chart showing survey results (N = 55).

It was noted that the students liked the hands-on and self-discovery approach, although at times some frustration was also noted. Once a demonstration was given there was only an interest to learn by themselves, back up when required by a Teaching Assistant's advice. The traditional view of CFD is that it has a steep learning curve, but with a structures CFD interface and with limited depth imposed it has been demonstrated that the gradient of the curve can be brought to an acceptable level.

Of course, during the skills training at this level, no real mention was made of code development, as the purpose was to develop users of the code only. This can be remedied by a later course which improves the student as a user and starts showing ways of writing new code for special conditions. Actually the software used here has a module which can translate simple instructions into FORTRAN. The concept to represent this software package or any other package as a black box should be remedied as soon as possible in later courses.

CONCLUDING REMARKS

This paper has described the use and efficacy of integrating computational fluid dynamics into a traditional fluid mechanics course. The controlled experiment has shown that the inclusion of CFD laboratories gave students a better appreciation of fluid mechanics in general and the students gained better knowledge of simple concepts. However, the inclusion of CFD laboratories had a detrimental effect on interest when compared to the purely experimental control group and the control group also did better when considering the more difficult aspects of the course. It was found from the study of student knowledge and skill outcomes for the CFD interface that the students could cope with CFD reasonably well, provided the subject is introduced with care. One of the main reasons for the inclusion of CFD was to contribute to the teaching of professional practice skills to intermediate level undergraduate students. It was found that the interface design does provide students with hands-on experience, gained through an interactive and user-friendly environment, and encourages student self-learning. It was noted from the survey that the students liked the hands-on and self-discovery approach, although at times some frustration was also noted.

REFERENCES

1. S. Aradag, K. Cohen, C. A. Seaer, T. Mc Laughlin, 2009Integration of Computations and Experiments for Flow Control Research with Undergraduate Students, Computer Applications in Engineering Education, 171111
2. D. Agonafer, D. G. Liao, B. Spalding, 2008The LVEL turbulence Model for Conjugate Heat Transfer at Low Reynolds Numbers, Concentration, Heat and Momentum Ltd., London, UK.
3. D. N. Assanis, J. B. Heywood, 1986Development and Use of a Computer Simulation of Turbocompounded Diesel System for Engine Performance and Component Heat Transfer Studies, SAE Transactions, 2451476
4. J. O. Campbell, J. R. Bourne, P. J. Mosterman, M. Nahni, R. Rassai, A. J. Brodersen, M. Dawant, 2004Cost-effective Distributed Learning with Electronics Labs, Journal of Asynchronous Learning Networks, 83510

5. J. Cohen, 1988Statistical Power Analysis for the Behavioral Sciences (2nd ed.). Hillsdale, NJ: Lawrence Earlbaum Associates.
6. W. J. Devenport, J. A. Schetz, 1998Boundary Layer Codes for Students in Java, Proceedings of FEDSM'98, ASME Fluids Engineering Division, Summer Meeting, Washington DC, June, 2125
7. L. D. Feisel, A. J. Rosa, 2002The Role of the Laboratory in Undergraduate Engineering Education, Journal of Engineering Education, 911121130
8. J. H. Ferziger, M. Peric, 1996Computational Methods for Fluid Dynamics, Springer, New York.
9. E. Gamliel, L. Davidovitz, 2005Online Versus Traditional Teaching Evaluation: Mode Can Matter, Assessment & Evaluation in Higher Education, 306581592
10. T. C. Hung, S. K. Wang, S. W. . Tai, C. T. Hung, 2005An Innovative Improvement in Engineering Learning System using Computational Fluid Dynamics Concept, Computer Applications in Engineering Education, 134306315
11. C. J. Keller, N. D. Finkelstein, K. K. Perkins, S. J. Pollock, 2007Assessing the Effectiveness of a Computer Simulation in Introductory Undergraduate Environments, AIP Conference Proceedings, 883121124
12. R. Kelsey, 2001Brownfield Action: An Integrated Environmental Science Simulation Experience for Undergraduates, Annual Proceedings of Selected Research and Development, Atlanta, GA, Nov. 812
13. R. D. La Roche, B. J. Hutchings, R. Muralikrishnan, (2002, Lab. Flow, Fluid. Computational, . C. F. D. Dynamics, for. Framework, Education. Undergraduate, of. Proceedings, 20. the, A. S. E. E. S. E. F. I. T. U. B. Colloquium, Oct. Berlin, 14
14. B. E. Launder, B. Spalding, 1972Mathematical Models of Turbulence, Academic Press. London.
15. R. Likert, 1932A Technique for the Measurement of Attitude, Archives of Psychology, 22140
16. U. Lucas, 1997Active Learning and Accounting Educators, Accounting Education, 6189190
17. Nulty, D.D., The Adequacy of Response Rates to Outline and Paper Surveys: What can be Done? Assessment & Evaluation in Higher Education, Vol.3301314

18. A. Patil, L. Mann, P. Howard, F. Martin, 2009Assessment of Hands-On Activities to Enhance Students' Learning in the First Year Engineering Skills Course, 20th Australasian Association for Engineering Education Conference, Univ. Of Adelaide, 6-9 Dec., 286292
19. D. Pfahl, O. Laitenberger, G. Ruhe, J. Dorsch, T. Krivobokova, 2004Evaluating the Learning Effectiveness of Using Simulations in Software Project Management Education: Results from a Twice Replicated Experiment, Information & Software Technology, 46127147
20. R. A. Pieritz, R. Mendes, Sila. R. F. A. F. Da, C. R. Maliska, 2004CFD: An educational Software Package for CFD Analysis and Design, Computer Applications in Engineering Education, 1212030
21. X. Qian, R. Tinker, 2006Molecular Dynamics Simulations of Chemical Reactions for Use in Education, Journal of Chemical Education, 8117790
22. J. W. Ray, W. R. Shadish, (1996, How Interchangeable are Different Estimators of Effect Size? Journal of Consulting and Clinical Psychology, Vol. 64, pp.1316-1325. (see also: (1998Correction to Ray and Shadish, Journal of Consulting and Clinical Psychology, 66532
23. G. Rozza, D. B. P. Huynh, N. C. Nguyen, A. T. Patera, (2009, Real, Real-Time Reliable Simulation of Heat Transfer Phenomena, ASME-American Society of Mechanical Engineers- Heat Transfer Summer Conference, Paper HT200988212San Franscisco, USA, July 19-23.
24. F. Stern, T. Xing, D. B. Yarbrough, A. Rothmayer, G. Rajagopalan, S. P. Otta, D. Caughey, R. Bhaskaran, S. Smith, Hutchings, B. , S. Moeykens, 2006Hands-On CFD Educational Interface for Engineering Courses and Laboratories, Journal of Engineering Education, 9516383
25. J. Tu, G. H. Yeoh, C. Liu, 2007Computational Fluid Dynamics- A Practical Approach, Elsevier, Amsterdam.
26. H. Zheng, J. M. Keith, 2003Web-Based Instructional Tools for Heat and Mass Transfer, ASEE Conference Proceedings, June.

CITATION

Desmond Adair (2012). Incorporation of Computational Fluid Dynamics into a Fluid Mechanics Curriculum, Advances in Modeling of Fluid Dynamics, Dr. Chaoqun Liu (Ed.), ISBN: 978-953-51-0834-4, InTech, DOI: 10.5772/39217.

CHAPTER 10

Computer Modelling of Automobile Fires

LadislavHalada, Peter Weisenpacher and Jan Glasa

[1]Institute of Informatics, Slovak Academy of Sciences, Slovakia

INTRODUCTION

The use of the *CFD (Computer Fluid Dynamics)* theory and its practical knowledge has become widespread in such academic disciplines as aerodynamics, fluid dynamics, combustion engineering and other fields. However, in the disciplines, which examine the ongoing processes in larger sizes, CFD was applied during the last decades only. One of such discipline is a spread of fire. Fire processes are a very complicated and complex phenomenon consisting of combustion, radiation, turbulence, fluid dynamics and other physical and chemical processes. A good knowledge about complex phenomena and processes occurring during fire in different environments is a significant component of fire safety. As fire itself is very complicated phenomenon, interdisciplinary approach to the problem is required.

This is alsoone of the reasons of a gradual development of programs in this area. In addition, large dynamics ofburning processes requires high spatial resolutionforsufficientlyaccurate calculation and also very smalltime step of calculation (thousandths of second) resulting in high demands oncomputer memory andhigh performanceprocessors.

Knowledge of and experience with the combustion processes were achieved mostly by experiments in the past. This form of investigation is quite difficult not only with regard to financial aspects but also from the point of view of variability of the input data related to burning.

In recent years, computer simulation of fire is used as an economically least expensive method to obtain the knowledge about ongoing fire processes and their visualization. This approach is especially valid in the case of fires in tunnels, car parks and buildings because full-scale fire experiments in such structures could cause serious damages of material and technical equipment. Nowadays, monitoring the development of processes in fire environment allows to achieve relatively good knowledge about the dynamics of liquids and gases. Existing software tools provided by CFD simulation also enable to visualize their development.

In the literature, computer simulation of fire was firstly formulated in the seventies as *zone models* and later as *multi-zone models* [1, 2, 3, 4]. In these models, the fire area is divided into separate fire areas (zones) so that each of these ongoing processes has been settled. The theoretical basis of these methods comprises the laws of conservation of mass and energy. The whole space is divided into two spatially homegeneous zones: warmer upper volume containing heat and smoke, and lower part significantly less affected by heat and smoke.For each zone, mass and energy balances are enforced with additional models describing other physical processes such as fire plumes, flows through doors, windows and other vents, radiative and convective heat transfer and solid fuel pyrolysis. Relative physical and computational simplicity of the zone models has led to their widespread use in the analysis of compartment fires scenarios. However, for certain fire scenarios, more detailed spatial distributions of physical properties are required.

Computational fluid dynamics (CFD) models were introduced in the nineties and have reached significant development and relatively widespread use in various fields of human activity. The rapid growth of computing power and advances in CFD have led to the development of CFD based field models based on solving the Reynolds-averaged form of the Navier-Stokes equations. The use of CFD models has allowed the description of fires in complex geometries incorporating a wide variety of physical phenomena related to fire.

The simplified equations developed by Rehm and Baum [5], referred to in the combustion research community as low Mach number combustion equations, describe low speed motion of a gas driven by chemical heat release and buoyancy forces [6]. These equations are solved numerically by dividing the physical space where the fire is to be simulated into a large number of rectangular cells. Within each cell, gas velocity, temperature, etc., are assumed to be uniform changing only with time. The accuracy with which the fire dynamics can be simulated depends on the number of cells that can be incorporated into the simulation. This number is ultimately limited by the computing power available. Nowadays, single processor computers limit the number of such cells to at most a few millions. Parallel processing can be used to extend this range to some extent, but the range of length scales that need to be accounted for if all relevant fire processes are to be simulated is roughly 10^4 to 10^5 because combustion processes take place at length scales of 1 mm or less, while the length scales associated with building or automobile fires are of the order of tens of meters/centimeters.

Several advanced systems intended for simulation of combustion processes have been developed. CFX, SMARTFIRE and FDS [7, 8, 9] and others provide alternative models which may offer a good performance. We use the Fire Dynamics Simulator (FDS) system [9], whose first version was developed in 2000 by the NIST (National Institute of Standards and Technology, USA). At present, significantly better FDS version 5.5 is available already. The development of both CFD software systems and their program modules will certainly continue. It is obvious that a careful verification and validation of these systems will continue in the future to enhance their quality and reliability.

BASIC EQUATIONS OF THE FDS MODEL

FDS solves a form of conservation equations for low speed, thermally driven flow. Smoke and heat transfer from fires is the main concern of this system, which also includes the thermal radiation, pyrolysis, combustion of pyrolysis products, flame spread and fire suppression by sprinklers. The basic set of the conservation mass, species, momentum and energy equations are as follows [9]:

$$\frac{\partial \rho}{\partial t} + \nabla . \rho \mathbf{u} = \dot{m}_b^{'''}$$
$$\frac{\partial}{\partial t}(\rho Y_\alpha) + \nabla . \rho Y_\alpha \mathbf{u} = \nabla . \rho D_\alpha \nabla Y_\alpha + \dot{m}_\alpha^{'''} + \dot{m}_{b,\alpha}^{'''} \quad (1)$$

$$\frac{\partial}{\partial t}(\rho \mathbf{u}) + \nabla \rho \mathbf{u}\mathbf{u} + \nabla p = \rho \mathbf{g} + \mathbf{f}_b + \nabla \tau_{ij}$$
$$\frac{\partial}{\partial t}(\rho h_s) + \nabla . \rho h_s \mathbf{u} = \frac{Dp}{Dt} + \dot{q}^{'''} - \dot{q}_b^{'''} - \nabla . \dot{\mathbf{q}}^{''} + \varepsilon, \quad (2)$$

where $\dot{m}_b^{'''} = \sum_\alpha \dot{m}_{b,\alpha}^{'''}$ is the production rate of species by evaporating droplets or particles; ρ is the density; u=(u,v,w) is the velocity vector; Y_α, D_α,and $\dot{m}_{b,\alpha}^{'''}$ are the mass fraction, the diffusion coefficient, and the mass production rate of α-th species per unit volume, respectively; p is the pressure; f_b is the external force vector; τ_{ij} is the viscous stress tensor; h_s is the sensible enthalpy; the term q''' is the heat release rate per unit volume from a chemical reaction and $\dot{q}_b^{'''}$ is the energy transferred to the evaporating droplets; and the term q''represents the conductive and radiative heat fluxes. Note that the use of the material derivative D( )/Dt= ∂ ( )/∂t + u.grad ( ) holds in the last equation.

Other two equations, the pressure equation and the equation of state,

$$\nabla^2 H = -\frac{\partial}{\partial t}(\nabla \mathbf{u}) - \nabla . F \quad \text{and} \quad p = \frac{\rho R T}{W} \quad (3)$$

are added to the previous four equations. The pressure equation is obtained applying the divergence on the momentum equation. In this equation, the value H represents the total pressure divided by the density. R is the universal gas constant, T is the temperature and W is the molecular weight of the gas mixture.

Thus, we have the set of six equations for six unknowns, which are functions of three spatial dimensions and time: the density ρ, three components of u = (u, v, w), the temperature T and the pressure p. These equations must be simplified in order to filter out sound waves, which are much faster than typical flow speed. The final numerical scheme is an explicit predictor-corrector finite difference scheme, which is second order accurate in space and time. The flow variables are updated in time using an explicit second-order Runge-Kutta scheme.

Boundary conditions are prescribed on walls and vents. All input data for simulation are required in the form of a text file in prescribed format, which describes the coordinate system, geometry of domain and its location in given coordinates, mesh resolution obstacles, boundary conditions, material properties and other different simulationparameters. An important limitation of the system is that the domain should be rectilinear, conforming with the underlying grid. The domain is filled with rectangular obstructions representing real objects, which can burn, heat up, conduct heat, etc. Simulation outputs include quantities for gas phase (temperature, velocity, species concentration, visibility, pressure, heat release rate per unit volume, etc.), for solid surfaces (temperature, heat flux, burning rate, etc.), as well as global quantities (total heat release rate, mass and energy fluxes through openings, etc.). These outputs are saved during simulation with the desired format for visualization and can be visualized by the Smokeview program.

As it is mentioned in [9], the overall computation can either be treated as Direct Numerical Simulation (DNS), in which the dissipative terms are computed directly, or as Large Eddy Simulation (LES), in which large-scale eddies are computed directly and subgrid-scale dissipative processes are modelled. The numerical algorithm is designed so that LES becomes DNS as grid is refined. The numericalschemes used for the solution of all equations is completely described in [9].

All FDS calculations must be performed within a domain consisting of rectilinear meshes divided into rectangular cells. These cells can be either uniform in size or stretched,fulfillingthe requirements of finite difference numerical scheme used in FDS. Their number depends on the desired resolution of fire scenario. As the FDS numerical scheme uses Fast Fourier Transforms (FFTs) in the y and z directions, the second and third mesh dimensions should each be of the form $2^l \times 3^m \times 5^n$ where *l*, *m*and*n* are integers.

In this chapter, we focus on computer simulation of:

- automobile engine compartment fire
- automobile passenger and/or luggage compartment fire
- fire spread from burning automobile onto a near standing automobile.

AUTOMOBILE FIRE EXPERIMENTS

Automobile fires, causing enormous losses of property and lives as well as large environmental damages, have become a significant phenomenon, highly injurious to public. They occur in different surroundings, both in open air (on roads, on open parking areas, in vicinity to forest) and in closed or semi-closed structures (in underground car parks, in garages, in tunnels). Such different types of fire show own specific behaviour and the way of spread. Therefore, fire investigation has to take into account various conditions affecting the fire development and serious research is of great importance.

Numerous papers dealing with different aspects of automobile fire safety have appeared and several advanced computer fire simulation systems have been developed. Such systems are capable to model, simulate and visualize the spread of flames and smoke and the fire behaviour, and even to estimate environmental damages caused by fire in various conditions.

Full-scale automobile fire experiments belong to the important means which are necessary for improvement of automobile safety. They provide information about overall automobile fire behaviour which may be difficult to extrapolate from laboratory testing of the automobile components. They are usually expensive to carry out, leading to short experimental series consisting of a few or even only one experiment. Typically they are extensively instrumented in order to obtain as much information as possible from one experiment. They require special experimental facilities because of great size of automobile fires leading to damage of real structures. Several experiments on automobile fires had been carried out under different conditions (open, closed, or semi-closed car parks) and reported in the literature.

The measurements in older experiments [10, 11, 12] were restricted to gas and structure surface temperatures in the vicinity of burning car. During three full-scale experiments reported in [13], the heat release rate, mass loss and mass loss rate, heat flux, carbon monoxide, carbon dioxide and smoke production rate and gas temperatures above the automobile, and temperatures inside the automobile were determined as a function of time. Two full-scale fire experiments in a tunnel shuttle wagon [14] were carried out. A series of 10 experiments (5 with one and 5 with two automobiles) with a ceiling above the automobile and spread of fire from one automobile to another [15] were performed, collecting the combustion products and oxygen consumption where rate of heat release was measured using oxygen consumption calorimetry. A series of 10 experiments (3 experiments with a single automobile, 6 with two automobiles and 1 with three automobiles parked next to each other) were carried out [16, 17, 18], where heat release rate was measured using oxygen consumption calorimetry. Two experiments with three latest generation automobiles parked in an open car park were carried out [19], where gas and steel temperatures and vertical displacements of steel beams were reported. One experiment in car park (3 automobiles in 0.5 and 0.7 m distances) was performed [20]. An experiment with automobile fire in a four-storey car park was performed [21], measuring temperatures, displacement and strain of car park structures. Four full-scale automobile fire experiments were carried out [22], igniting in different parts of automobile and measuring mass loss, mass loss rate and temperature curves.

In 2009, a series of three full-scale automobile fire experiments with 4 cars in open air in the testing facilities of Fire Protection College of the Ministry of Interior of Slovak Republic in PovazskyChlmec (Slovakia) and one full-scale automobile fire experiment with 2 cars in 0.5 m distance in the experimental tunnel of Scientific Research Coal Institute in Stramberk (Czech Republic) were carried out [23, 24, 25]. The primary objective of these experiments was to measure gas and surface temperatures (on the surface, above and inside the automobile and engine compartment) to gather data for validation of computer simulations of tested automobile fire scenarios. The main aim of the experiments was to obtain better knowledge about burning of a single automobile, to determine most combustible materials contributing to fire and to study the spread of fire from one vehicle to another.On the basis of these experiments we have consequently made computer simulations.

AUTOMOBILE ENGINE COMPARTMENT FIRE EXPERIMENT AND SIMULATIONS

This type of automobile fire has specific fire behaviour and nowadays it belongs to the most frequent vehicle fires [25, 26, 27, 28]. According to our knowledge, the spread of fire inside the engine compartment has not been analysed by computer fire simulator and published yet. We believe that such type of fire analysis and consequent computer simulation is very important. It provides more realistic fire spread in vicinity of the automobile and therefore more realistic input data for computer fire simulations in car parks, garages, or tunnels. We conjecture that ignition process for car fire simulations in such places simulated by a pool with burning gasoline (for example) is less realistic because it produces constant heat release rate.

The reported automobile fire experiments were carried out on automobiles of the same category. In the sequel, we focus on the first fire experiments from the reported series of full-scale automobile fire experiments. The first onewas carried out on Audi 80 Quattro with just slightly distorted bodywork and missing front lights parked in open air on a concrete surface (Fig. 1). All doors and windows of the automobile were closed during the experiment. Gas temperatures inside the engine compartment (on the upper part of the engine block), inside the passenger compartment (in front part on the dashboard) and above the automobile (at 1.5 m height) were measured by thermocouples with 10 s recording interval. Temperature on the engine compartment lid surface was recorded by infra-red camera placed in front of the automobile at 3.5 m height and at 6 m distance from the automobile. The fire behaviour observations were recorded by digital cameras. Basic meteorological data (exterior temperature, wind speed, atmospheric pressure and others) were also collected.

Figure 1: Automobile with thermocouples prepared for the experiment.

The fire was ignited by burning of a small amount of gasoline (about 5 ml) poured onto a small cloth, which was placed in the engine compartment under the rubber tube (Figures 2 and 3) imitating a frequent failure in the engine compartment. Immediately after the fire ignition, the engine compartment lid was closed. Sparse smoke above the lid appeared very soon and became denser gradually. During the first three minutes, smoke from the interstices at right edges of the lid and on the lid below the front window was prevailing. At the 4th minute, the smoke leaking out from the holes at the place of missing front lights became more intensive. At the 7th minute, weak flames appeared in the hole at the place of the missing left front light. Between the 7th and 8th minute of fire, flames appeared on the surface on the left side of the lid (varnish ignition). After 12 minutes, the fire was extinguished and inspection started to determine the components degraded most by the fire, and to detect the most flammable materials in the engine compartment.

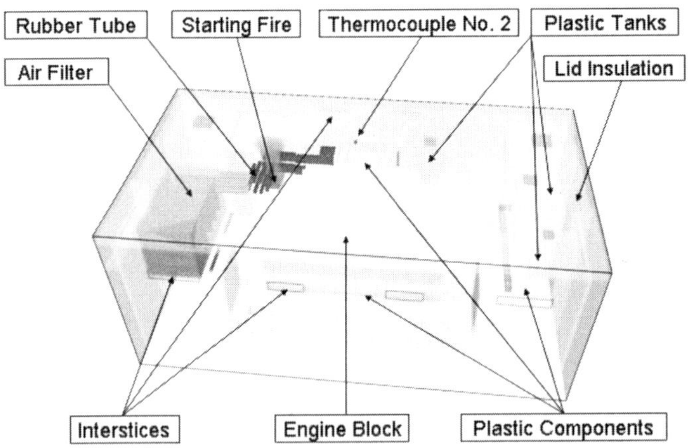

Figure 2: Scheme of automobile engine compartment.

Figure 3: Engine compartment before (left) and after (right) the fire experiment.

The most flammable objects in the engine compartment, completely or almost completely burnt during the first 12 minutes of burning, were: the rubber tube (above the fire ignition source), 3 plastic tanks (in the right back part of the engine compartment), air filter box with its paper content (on the left side of the engine compartment), and several other plastic components (Figures 2 and 3).It was confirmed that these components contribute most to fire in engine compartment. The inspection also showed significant degradation of several electric cables and other hoses, partially auto battery and partially molten parts of engine block, and other components afflicted by fire. However, the contribution of these components to fire seems to be small. Fig. 4 shows that the temperature recorded by the thermocouple in the automobile engine compartment reached as many as 900 °C.

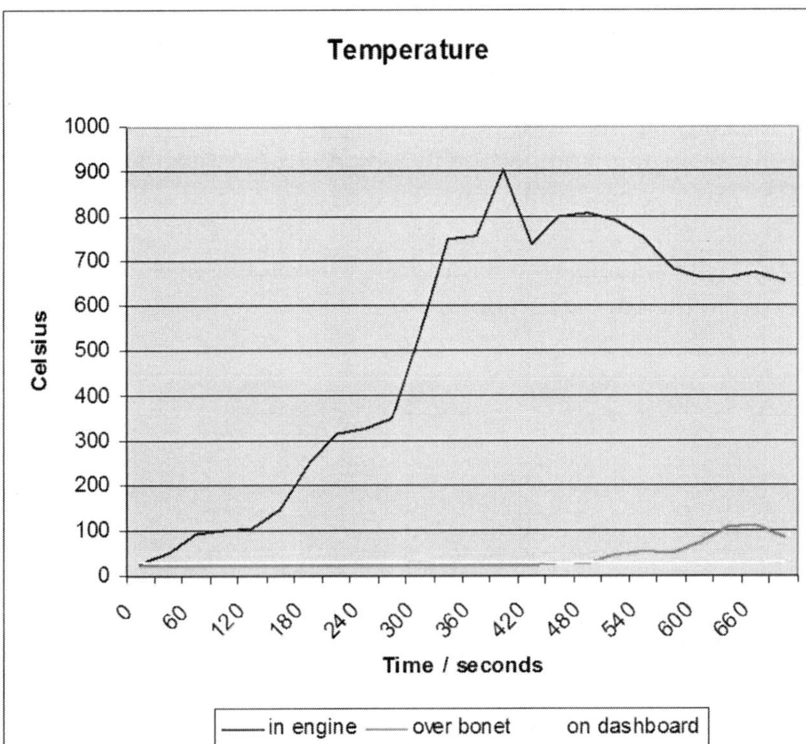

Figure 4: Temperature in the engine compartment, over the bonnet and on the dashboard.

The next automobile fire experiment modelled a possible real situation of fire occured during car driving. The driver observed fuming, stopped the car after some hesitation, and decided to extinguish the fire in engine compartment through slightly ajar lid by portable car fire extinguisher. During this experiment,a technical failure was confirmed.

Fire fighters, who should have fought the fire, could not open the engine compartment due to the melting of the lock control mechanism of the engine compartment lid. It was even not possible to open the engine compartment by heaver and plate shears had to be used (see Fig. 5). For this reason, the fire fighting could start with a substantial delay only. Thus, it can be concluded that cars with such plastic control lock mechanism of engine compartment lid may not be extinguished by portable car fire extinguisher due to inaccessibility of the space where the combustion takes place.

The temperature peak over the bonnet comes earlier than that in engine compartment. Probable reason of that is insufficient oxygen supply in engine compartment, which suppressed the fire spread inside the engine compartment and caused that the thermocouple in the engine compartment was not exposed to direct plume. Instead, the sealing compound burnt down and gaseous fuel was created in the engine compartment.

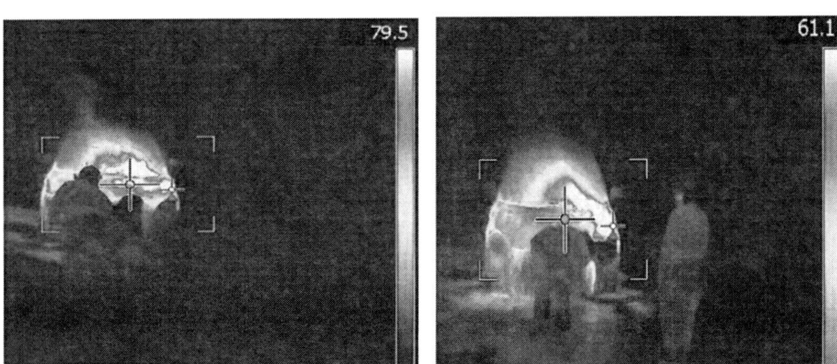

Figure 5: Destruction of the plastic control lock mechanism of engine compartment lid.

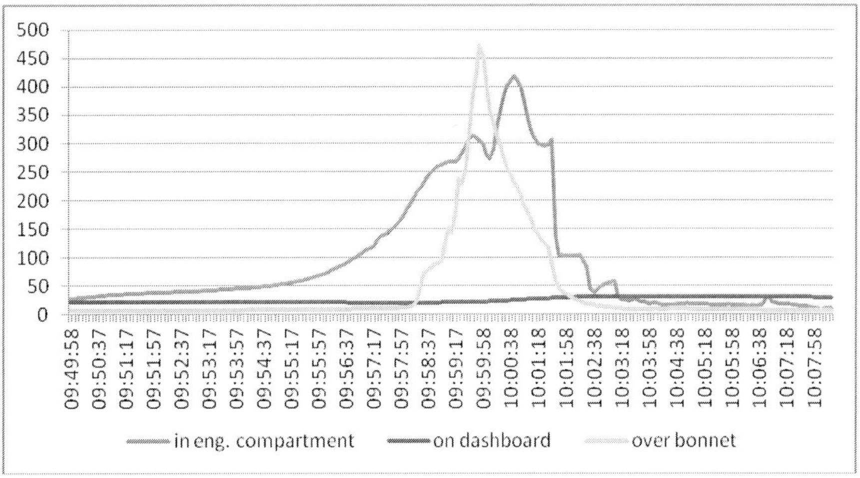

Figure 6: Temperature in the engine compartment, over the bonnet and on the dashboard.

Computer Simulation of Engine Compartment Fire

Simulation of automobile engine compartment fire belongs to the most complex FDS simulation problems. Due to the extremely complex geometry of the burning space and objects inside the engine compartment, which affect the fire development and then have to be accurately captured, such a simulation requires very fine mesh resolution and therefore significant computational power for calculation. That is why the design of engine compartment geometry and of the components in the input FDS file is of great importance. In order to properly model the engine compartment as well as all relevant flammable components in its interior, corresponding input geometry of simulated space was elaborated using available 3D scans, as well as direct measurements of distances and proportions of detected flammable components (done before the start of experiment). The simulation domain includes all flammable components mentioned above and other plastic and metallic components which influence the direction of fire and smoke spread in time (see Fig. 2).

Engine compartment fire simulation is just the situation in which certain geometric features of engine compartment components do not conform to rectangular mesh, and have to be represented in a

different way (electrical cables, tubes, hoses, etc.). The shape of plastic tanks and air filter can be captured almost realistically. However, the rubber tube (small thickness and cylindrical shape) has to be represented by a cluster of thin stripes where the total mass of stripes is equal to the mass of the tube itself. Moreover, the surface to volume ratio of the stripes is the same as that of the tube. Both these parameters, which are crucial for heating up and burning of material, were maintained. By this way some other components, such as paper interior of the air filter box, were modelled. Several other components, which did not appear to be critical for heat transfer and burning (e.g. small plastic caps of tanks too small to be properly captured), were represented by plastic or metallic slabs placed at appropriate places in the simulation space.
Proper determination of material properties (physical parameters describing burning properties of materials) in the engine compartment was another essential task to be solved. There were four prevailing types of materials identified: aluminium alloy (metallic components), rubber (rubber tube), paper (air filter interior) and mixture of polyethylene (plastic components). Some material parameters for automobile varnish were estimated (e.g. the heat release rate per unit area) and some of them were derived from observations (e.g. the ignition temperature was determined from recorded infra-red camera observations) during the experiment. Most material parameters were determined by specialists from University in Zilina (Slovakia).

In the simulation, the fire ignition source (the small burning cloth placed on the engine block under the rubber tube) is represented by a burning surface with the dimensions of 4 x 4 cm and total heat release rate set to 2.1 kW for the period of 60 s.

Several measurement devices were defined in the simulation to provide proper output parameters describing the fire behaviour in time, such as gas temperature and velocity, surface temperature, oxygen concentration, etc.

Fds Simulation Results

For the simulation of the fire scenarios described above, we used a PC with 4-core processor Intel Q9550 with the frequency of 2.83 GHz, 8 GB of RAM and 1TB hard disk.

Two types of the engine compartment fire simulation are briefly presented in this section. They differ from each other:

- in the manner in which the computational domain was chosen
- by their computational requirements, the range of consideration of the automobile environment in fire scenario
- by the number and precision of output quantities provided.

For each case, dozens of simulations were performed in order to test and find proper values of significant parameters and appropriate location of interstices affecting the oxygen supply.

Simulation A

In this simulation, the computational domain includes the interior of engine compartment and 3 cm space above the engine compartment lid to show also the temperature distribution on the upper surface of the lid. The domain boundary conditions defining the heat transfer are given by the material properties of the bodywork (aluminium alloy) and are set 'OPEN' for the upper boundary of the 3 cm space above the engine compartment. The interstices in the bodywork (Figures 2, 3 and 7) are represented by narrow vents with boundary conditions 'OPEN', which allow convective gas transfer between the engine compartment and the automobile environment, not explicitly included in this simulation. They represent the holes at the place of missing front lights and the corresponding interstices at the front and back sides of the engine compartment and on the lid edges.

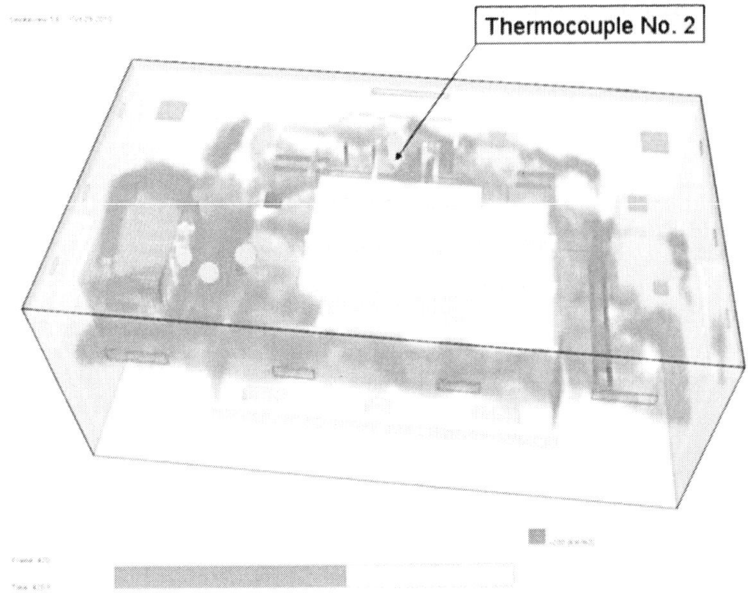

Figure 7: Simulation A in the 7th minute of burning. The orange colour volumes represent the volumes with HRR per unit volume values higher than 200 $kW.m^{-3}$.

The whole computational domain is represented by one computational mesh of 1 cm resolution with 503,712 cells which was assigned to one CPU core. Most of the dynamic processes of burning occurred in the left part of the mesh. The simulation of 720 s of fire required 207 hours of CPU time at Intel Q9550, 2.83 GHz CPU.

The main fire spread tendencies are similar to observations during the fire experiment. The simulated fire behaviour was as follows. At the 22nd second of fire, the rubber tube and the lid insulation layer above the burning cloth started to burn. At the 50th second of burning, flames were spread in the whole engine compartment. At that time, the total heat release rate (HRR) of about 15 - 20 kW was achieved. In rear part of the engine compartment, significant ascending air flow was created by the interstices located under the front window, which made the fire spread along the rubber tube to the rear part of the engine compartment.

This fire behaviour is in accordance with the observation of intensive smoke from the interstices below the front window in this phase of fire and with increase of the lid temperature detected by IR camera, as well as with increase of thermocouple No. 2 during the fire experiment (see Figures 4 and 8).

Meanwhile, the air filter was heated up and its paper content ignited, although its HRR was very low because of insufficient oxygen supply. Between the 200th and 300th second of simulation, the plastic components in the rear part of the engine compartment burnt away, which led to significant increase in air supply and caused temporary temperature increase at sensor No. 2 and to more intensive burning of the air filter. This behaviour is in accordance with the fire experiment, which showed strong increase of temperature of a part of the lid above the air filter in that phase of fire (see the surface temperature distribution on the lid shown in Fig. 8 as well as the peak in Fig. 4), as well as intensive smoke produced in the front part of the automobile. Although the range of colour schemes for surface temperatures in simulation and IR camera outputs in Fig. 8 are not the same, the temperatures show similar behaviour.

Heat and flames spread mainly along the air filter in the direction towards the hole at the place of missing left front light. At the 428th second of simulation (440th second in the fire experiment), the outer lid surface reached the temperature of 300°C and the varnish ignited. After the air filter burnt away, the plastic components in the front part of the engine compartment started to burn more intensively (flames in the holes at the place of missing front lights were observed during the experiment too). The maximal value of total HRR achieved was about 40-50 kW.

The maximal surface temperature on the lid achieved by simulation is well comparable to the values recorded by the IR camera. It can be seen in Fig. 8 that the lid surface temperature distribution in time is in good spatial and temporal accordance with IR camera records (see also Fig. 9). The time of varnish ignition on the lid surface corresponds with observations during the experiment.

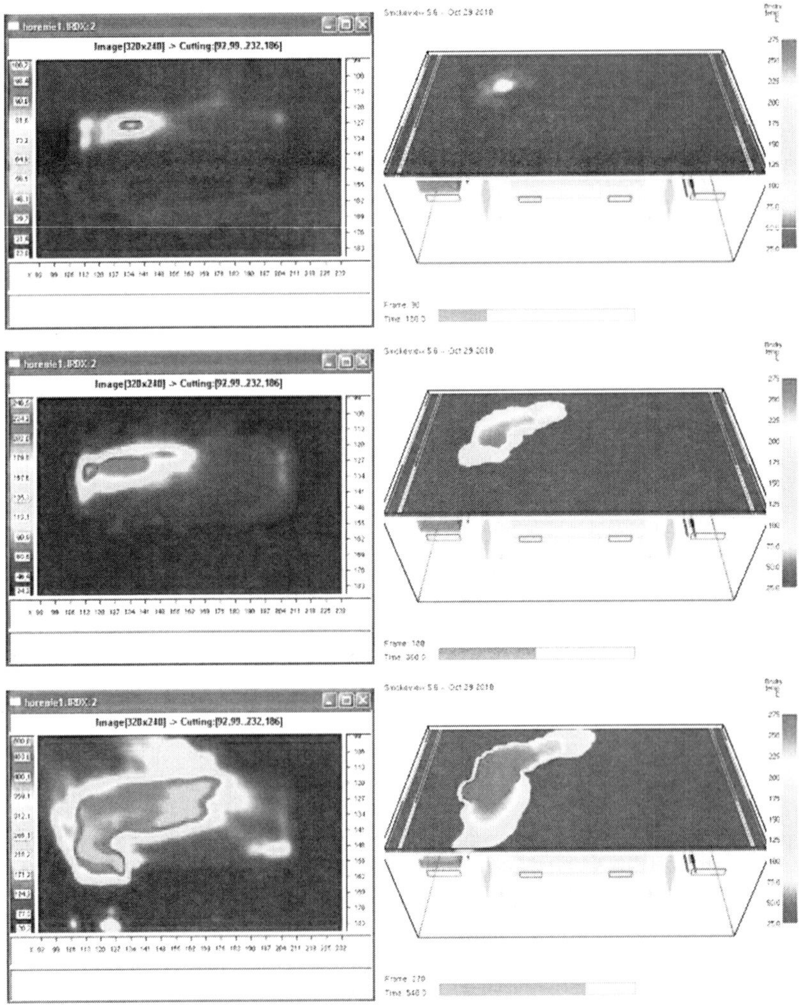

Figure 8: Surface temperature distribution on the engine lid: in IR camera output (left) and in simulation A (right) in the 3rd, 6th and 9th minute of burning.

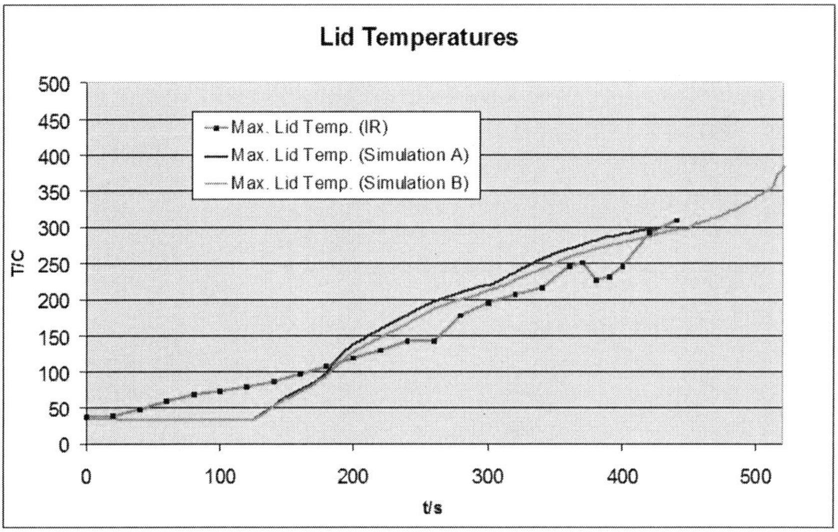

Figure 9: Maximal temperature of the lid surface in simulation and in the IR camera output.

Simulation B

In this case, the computational domain includes the engine compartment interior as well as additional space above the engine compartment lid providing information about the fire spread above the engine compartment (see Fig. 10). The domain boundary conditions defining the heat transfer are again given by the material properties of the bodywork (aluminium alloy) and 'OPEN' for the space above the engine compartment. The interstices in the bodywork are represented as before by vents with boundary conditions 'OPEN'.

The additional mesh placed above the lid includes the space, in which potential varnish fire occurs during the simulation. Its resolution decreased to 2 cm having 192,640 additional cells. The engine compartment itself was assigned to one computational mesh. Most dynamic processes of burning occurred inside the engine compartment, therefore, the impact of such division on the accuracy of computation was small.The additional space above the engine compartment lid did not influence the simulation of fire in the engine compartment interior much (see Fig. 9). However, the simulation provides more realistic visualization of fire above the lid (see Fig. 10). The advantage of case B in comparison with case A is that it provides further additional output quantities of fire outside the engine compartment (e.g. the HRR of varnish fire, temperature above the lid, etc.) at the cost of growth of computational requirements. The simulation of 700 s of fire required 200 hours of CPU time at Intel Q9550, 2.83 GHz CPU.

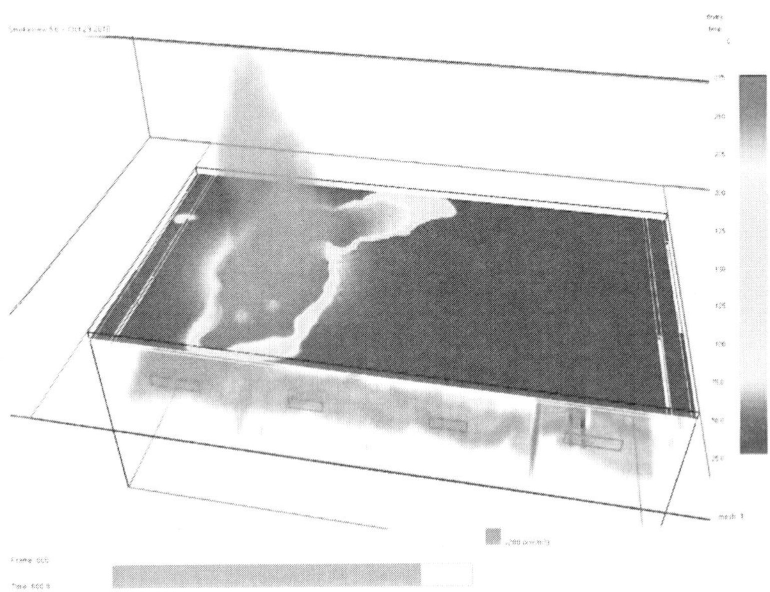

Figure 10: Simulation B in the 10th minute of burning. The orange volumes represent the ones with HRR per unit volume values higher than 200 kW.m^{-3}.

The main tendencies of the fire spread in this simulation are similar to the fire behaviour in simulation A and to observations during the fire experiment. The results of HRR of both simulation A and B are shown in Fig. 11.

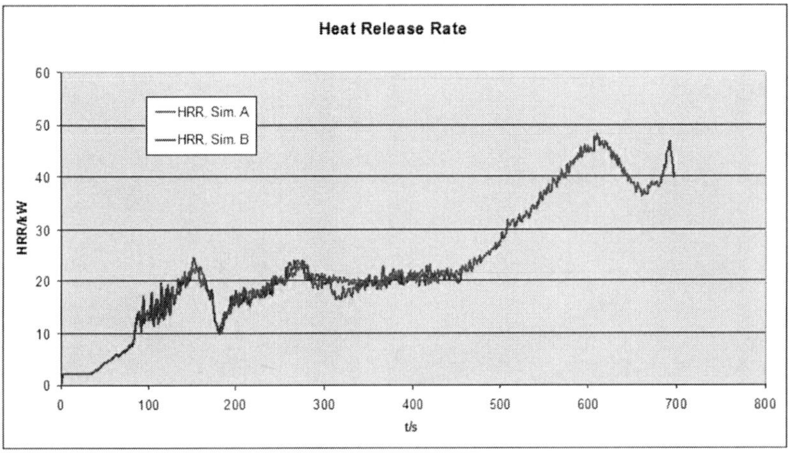

Figure 11: Heat release rate of fire in the simulations A and B.

AUTOMOBILE FIRES IN PASSENGER AND LUGGAGE COMPARTMENTS

Fires of automobile passenger and/or luggage compartment are not very frequent. However, if they occur, they can be very dangerous particularly in urban conglomeration. They often accelerate fast especially in sooner manufactured automobiles with more flammable car seat upholstery materials and/or flammable vehicle load. Such fires can cause large damages especially in the case, when the interior of automobile is sufficiently supplied with oxygen during fire (for example, if it flows through a broken window or ajar door).

In this section, we will describe the FDS simulations of automobile fires in passenger and luggage compartment related to one of three full-scale fire tests conducted during the third fire experiment carried out in Slovakia in 2009 (see Fig. 12).

Figure 12: Full-scale fire experiment: passenger compartment fire.

The fire was initiated on the back seat behind the driver seat. The left back window was open. The aim of the experiment was to monitor the course of passenger compartment fire and subsequent spread of fire from burning automobile onto a near-standing vehicle. Temperatures were measured by thermocouples on the engine block, over the engine compartment and in the middle of dashboard. Lateral air flow was imitated by a fan. The experiment showed a strong relation between the passenger compartment fire and the air supply via the open window. Substantial fire acceleration was observed in the moment of the windscreen destruction. High plumes and dense smoke appeared in the passenger compartment very soon in the same minute as the fire was ignited. The temperature in the passenger compartment reached the value of 750°C in the second minute of fire. In the 3rd minute of burning, the windscreen destruction and airbags explosion occurred followed by strong plumes. At the 4th minute of fire, a driving mirror burning was observed caused by over-jumping a burning piece of flammable material from burning automobile in consequence of the airbag explosion. We performed a simulation of the fire scenario described above. The geometry of the used automobile was constructed and flammable components in the passenger compartment were modelled by the upholstery and plastic volumes (see Fig. 13).

The real material properties were estimated, or gathered from tables. Boundary conditions 'OPEN' were used for the domain boundaries. For the simulation, one computational mesh of 3 cm mesh resolution consisting of 1,049,760 cells was used. The simulation of 600 s of fire required 130 hours of CPU time at Intel i7-990X CPU, 3.46 GHz CPU. Some typical phases of the simulated passenger compartment fire are demonstrated in Figures 14 and 15.

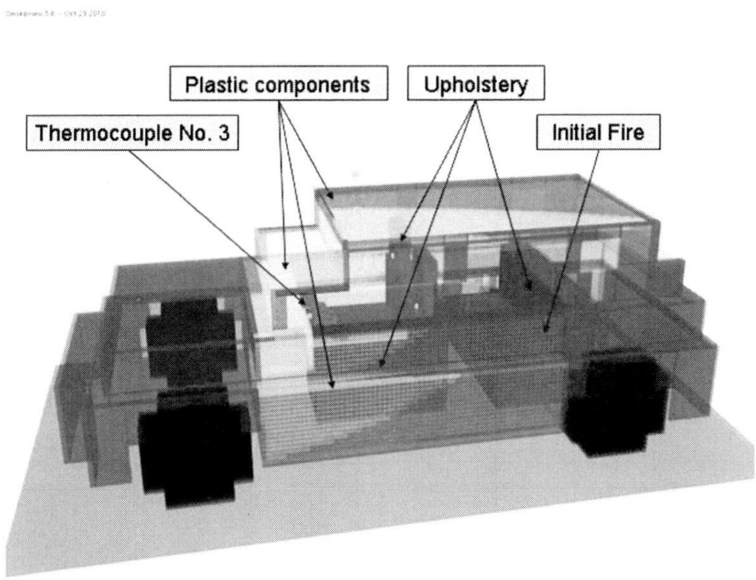

Figure 13: Scheme of automobile passenger compartment and used materials.

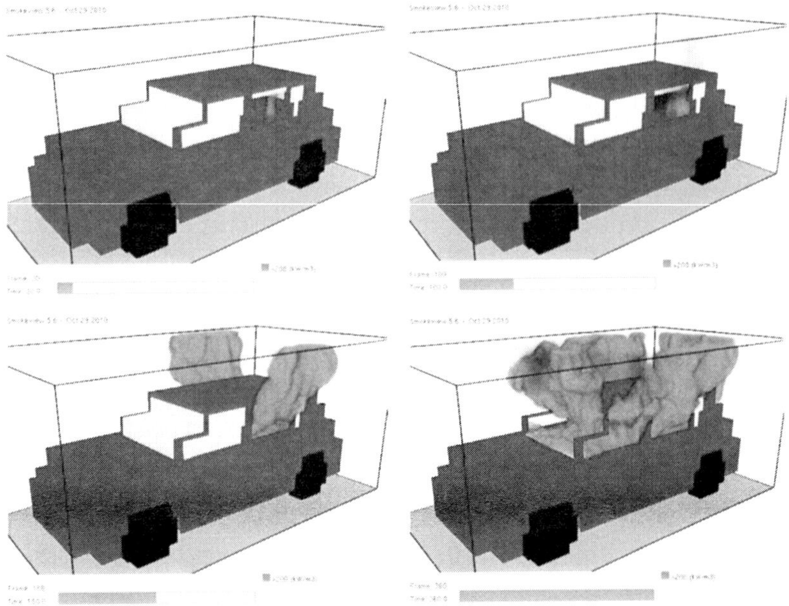

Figure 14: Simulation of passenger compartment fire in the 30th, 100th, 180th and 360th s of fire.

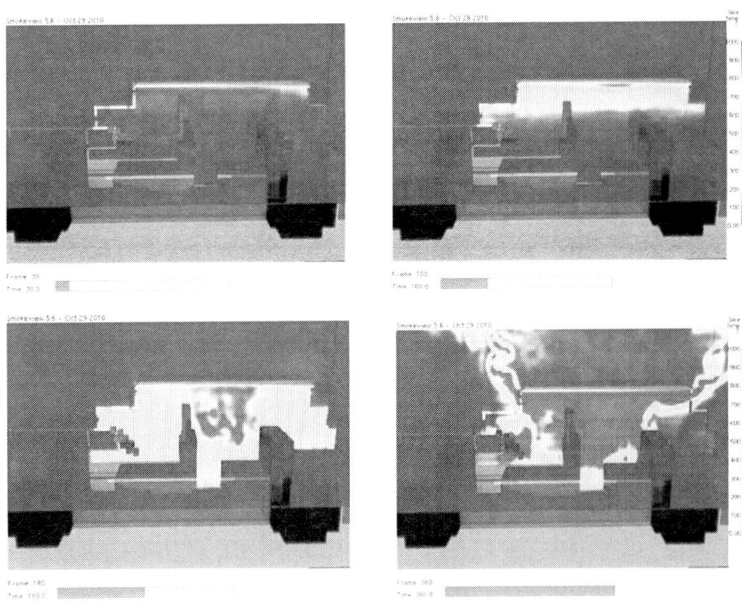

Figure 15: Simulation of passenger compartment fire: temperature cuts in the 30th, 100th, 180th and 360thsecond of fire.

The fire behaviour was as follows. The fire was ignited in the same place as in the fire experiment, on the back seat behind the driver seat. Significant air supply through open left back window enabled rapid spread of fire throughout the whole back seat and significant increase of HRR. Other windows were removed in the simulation, step by step by observations from the fire experiment. Thus, the air supply increased even more in the next phases of burning. After 115 s of simulation, the whole back part of the passenger compartment was ignited and the right back window was removed. After 240 s, all remaining windows were removed and the HRR reached the maximal value of about 4 MW.

Fig. 16 shows the HRRbehaviour during the fire, while temperature in passengercompartment and its comparison with the experimental values (thermocouple No.3) are shown in Fig. 17.
The simulation results are in good agreement with experimental observations, although the value of the maximum temperature in simulation is higher by about 200 °C. In order to make this simulation more accurate, some additional research of real upholstery and plastic components material properties is required. The "step" behaviour of the temperature curve in the simulation, which can be seen in Fig. 17, was caused by the sudden removal of windows. In the real fire, this removal was more gradual. Therefore, no such sudden increase occurred. Thorough examination of material properties of glass and its incorporation into the simulation can solve this problem.

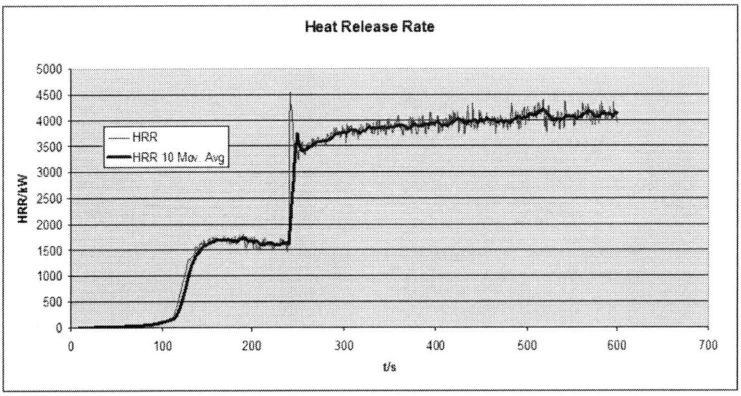

Figure 16: HRR of passenger compartment fire.

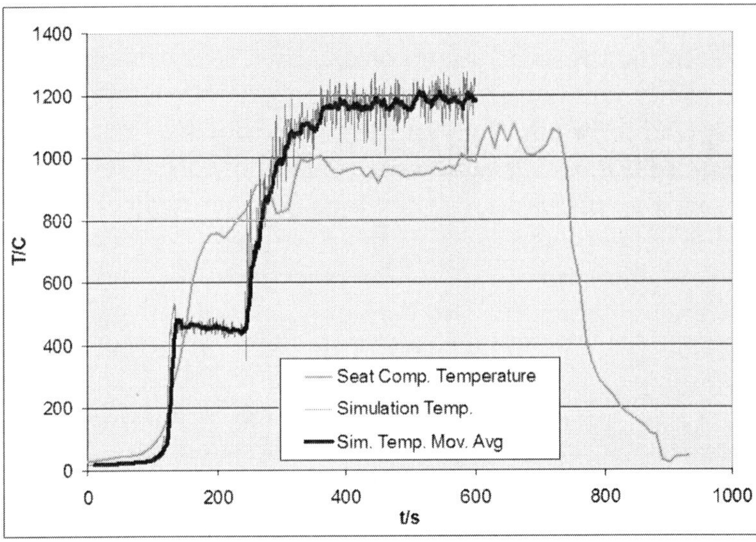

Figure 17: Temperature curves in the passenger compartment interior during the fire: simulation and experiment.

FDS SIMULATION OF FIRE SPREAD FROM BURNING AUTOMOBILE ONTO A NEAR-STANDING VEHICLE

In this section, the FDS simulations of automobile fire spread onto a near-standing automobile will be described which are related to one of three full-scale fire tests conducted during the fourth fire experiment carried out in Slovakia in 2009 (see Fig. 18).

Figure 18: Full-scale fire experiment: spread of fire onto near-standing automobile.

A simple case of the simulation of near-standing automobile ignition from burning vehicle is shown inFig. 19. The aim of the simulation was to investigate the influence of wind speed and distance between the automobiles upon automobile ignition. Three automobiles were used in the simulation, two of them are of the same type as in the previous simulation. The first automobile (left) modified the air flow profile only. The second (central) automobile with slightly different bodywork geometry represented a source of fire. The third automobile was exposed to the central vehicle fire. The air flow direction was chosen to accelerate the spread of flames from the second to the third automobile. The distance between the second and the third automobiles and air flow velocity varied from 15 cm to 75 cm and from 0 $m.s^{-1}$ to 3 $m.s^{-1}$, respectively. The total HRR of burning engine compartment was 787.5 kW (HRR per unit area was 500 $kW.m^{-2}$). In order to evaluate the impact of fire on the third automobile, a thermocouple was placed on the right side of its bodywork (Fig. 19).

The computational domain size was 810 x 540 x 225 cm with 5 cm computational mesh resolution. The total number of cells was equal to 787,320. Boundary conditions for the left and right domain boundaries (see the green surfaces in Fig. 19) are defined by the air flow velocities considered in the fire scenarios. Boundary conditions 'OPEN' are set for other domain boundaries. Intel i7-990X CPU, 3.46 GHz CPU was used for a series of simulations with different distance and velocity parameters. Duration of 100 s fire simulation was approximately 12 hours. Maximal simulation time was 600 s. By this way, it is possible to simulate and analyze the ignition time of near-standing automobile depending on different parameters in different conditions in open, semi-closed and closed areas for miscellaneous car categories.

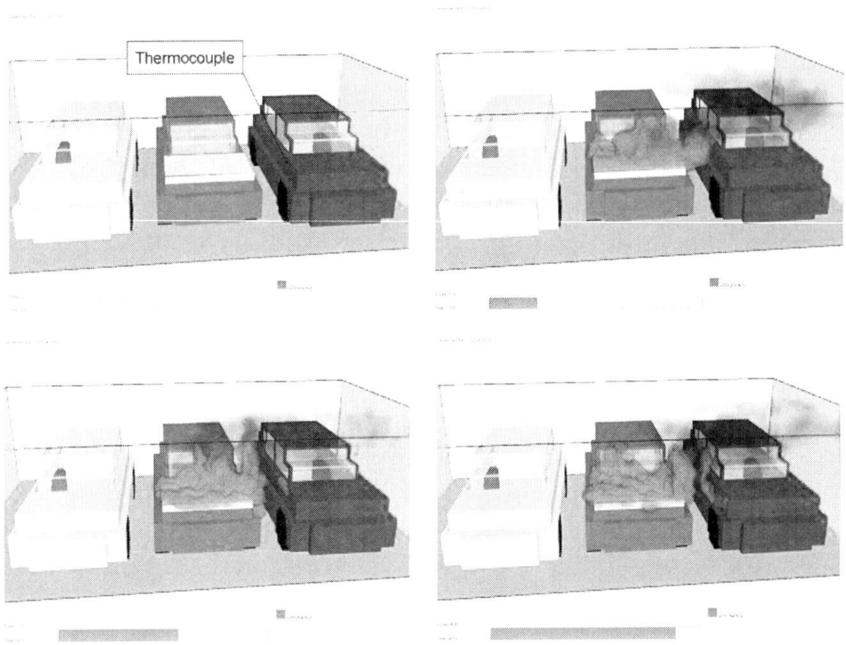

Figure 19: Simulation of near-standing automobile ignition: the case of 45 cm distance between automobiles and 2 $m.s^{-1}$air flow velocity in the 0th, 120th, 300th and 465th s of fire.

Some typical phases of the simulation with 45 cm distance between automobiles and 2 $m.s^{-1}$ air flow velocity are shown in Fig. 19. In the simulation, the ignition occurred at the 288th second of burning. In Table 1, the simulation results for different combinations of the chosen parameters are shown.

Table 1: Time of the third automobile ignition for different distance and airflow velocity values

	0 $m.s^{-1}$	**0.5 $m.s^{-1}$**	**1 $m.s^{-1}$**	**2 $m.s^{-1}$**	**3 $m.s^{-1}$**
15 cm	304 s	251 s	185 s	137 s	107 s
30 cm	384 s	327 s	254 s	193 s	167 s
45 cm	494 s	441 s	348 s	288 s	268 s
60 cm	-	-	474 s	407 s	389 s
75 cm	-	-	-	575 s	-

Fig. 20 shows the temperature behaviour at the thermocouple placed on the third automobile (see Fig. 19) for 2 m.s^{-1} air flow velocity and different distance values.

The simulation results confirmed the observations during the fire experiments conducted, that 60 cm distance between automobiles can be considered to be a reasonable safe distance between automobiles in car parks. For this value of the distance, ignition occurred only if the air flow velocity was above 1 m.s^{-1} and a relatively intensive fire lasted for almost 10 minutes. Such conditions are relatively rare in semi-closed or closed compartments.

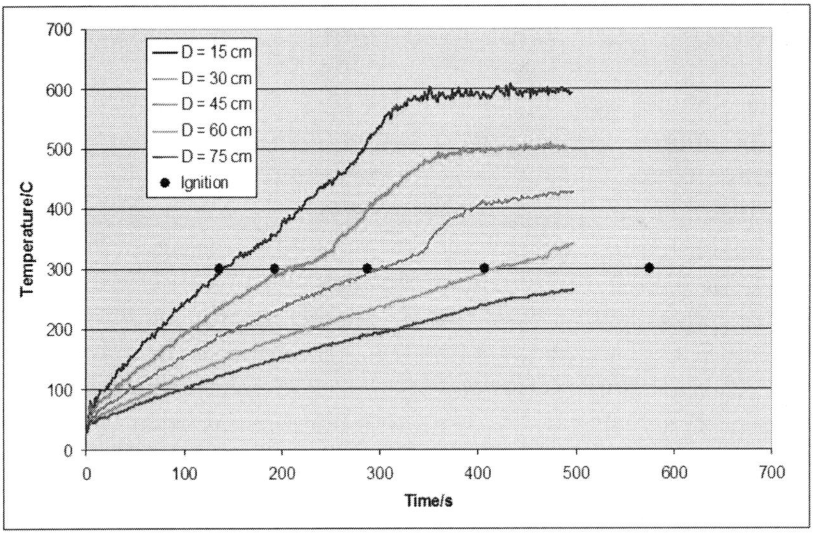

Figure 20: Temperature behaviour for 2 m.s^{-1} air flow velocity and different distance values. Black points represent the time of ignition of the third automobile observed in simulation. This time is not inevitably the same as the time when the temperature in control thermocouple reached the ignition temperature. The place of ignition was different in every simulation.

Another important observation is that a higher air flow velocity does not necessarily mean a higher probability of the ignition. For 75 cm distance and 3 m.s^{-1} air flow velocity, theignition did not occur, although for 3 m.s^{-1}velocity the third automobile was ignited.

In this case, a cooling effect of strong air flow and its influence on flame distraction probably prevailed and heating effect of burning vehicle was suppressed.

Simulations of this kind can be used for specific automobile fire scenarios expected in practice to evaluate the impact of fire, as scenarios parameters can be easily modified according to user requirements.

CONCLUSION

In this chapter, three simple cases of typical automobile fire safety problems arising in real traffic situations and their computer simulation are briefly described:

- automobile engine compartment fire
- automobile passenger and/or luggage compartment fire
- fire spread from burning automobile onto a near-standing automobile.

These simple fire scenarios were simulated by using advanced fire simulator FDS (Fire Dynamics Simulator) and validated byresults of the full-scale fire experiments in the testing facilities of Fire Protection College of the Ministry of Interior of Slovak Republic performed in 2009 in PovazskyChlmec(Slovakia). These fire types correspond to most frequent automobile fires associated with real situations on roads, parking areas, underground car parks, garages and road tunnels. During these experiments, several automobile fires have been performed to better understand the burning process itself and to collect proper fire parameters describing the fire behaviour for the purposes of verification of the computer FDS simulations of these fires.

The developed simulations indicate the ability of the used fire simulation system to model, simulate and visualize various practical fire scenarios and can help test the fire behaviour and various critical parameters which are important for automobile fire safety (such as e.g. critical distance between burning vehicle and near-standing automobile and/or other objects, ignition time of near-standingobjects afflicted by automobile fire or vice-versa, tunnel ventilation parameters in the fire regime, etc.). Computer simulation allows also to test the influence of flammable material properties on the fire behaviour for certain fire scenario. The simulation can provide us with information about the main fire spread tendencies, as well as about its quantitative characteristics like fire heat release rate, temporal temperature curves, etc. It allows to evaluate fire risk for the automobile environment and to test possible fire scenarios for concrete situations, which can help evaluate fire suppression options in further fire phases.

General agreement between the simulations presented and the fire experiment observations have been confirmed. However, several specific features can be seen when analyzing the suggested simulations:

- Size of flames from engine compartment interstices in initial phase of the simulation. They seem to be longer than observed. A possible reason of that phenomenon can be the visualization of hot gases coming out from interstices (HRR per unit volume of which exceeds 1199 $kW.m^{-3}$), which are visualized as "flames" but are still not recognized as "flames" by human eye. Moreover, the real interstices were a bit narrower and longer than the simulated ones, maintaining the realistic total leakage area for the air flow, and they have more intricate form (of cavity under the lid). This imperfect "geometry" of interstices following from the mesh limitations could not entirely capture more complicated burning in these small, but intricate spaces. Hot gas could interact with metal surface of the lid in more complicated way than the simple open vent in the simulation. Although this phenomenon was manifested in the simulation B by a bit longer "flames" coming out from the interstices, the simulation of burning inside the engine compartment was reliable.
- Limitations of models implemented in FDS concerning heat conduction modelling in solids probably affect the surface temperature distribution in the lid. Since the FDS does not simulate 3D heat transfer in the lid, the shape of isotherms on the lid is slightly different than in the IR pictures recorded during the fire experiment. However, this influence was of minor importance for the simulation reliability.

Despite the fact that the automobile engine compartment fire is a complex FDS simulation problem, the FDS simulator proved its applicability for modelling such complex fires. Although the exact geometry of objects in the simulated space is important, if the mesh resolution limitations prevent to capture them precisely, at least their surface to volume ratio and the total mass should be maintained. Air supply and proper size and position of interstices is crucial for the fire development; however, this information is usually not available and user should experiment with several simulations in order to find the proper solution. Next important problem is the determination of material properties where small differences in some values can change the fire behaviour qualitatively. Moreover, efficient realization and validation of fire simulations of large places (for example in a road tunnel) for fire safety purposes in parallel using parallel computers and distributed environments is also a challenging problem.

ACKNOWLEDGEMENT

This work was partially supported by Slovak Scientific Research Agencies VEGA(project VEGA 2/0216/10) and ASFEU (project OPVV CRISISITMS 26240220060). Authors would like to thank to plk. Ing.JaroslavFlachbart, PhD., the director of Fire Protection College of the Ministry of Interior of Slovak Republic in Zilinafor his kind help during the full-scale automobile fire experiments conducted in cooperation with the research team from the University of Zilina led by Prof.PavelPolednak, PhD.

REFERENCES

1. B. Zhao, 2001Numerical Modelling of Structural Behaviour of Open Car Parks under Natural Fire. Proceedings of the 9th Int. Conf. Interflam, 1September 2001), Edinburgh, Scotland, 3833930-95323-127-5
2. C. Steinert, 1999Feuerubersprung und Abbrandverhalten von PersonenkraftwagenTl. 1. Stuttgart: Fraunhofer IRB Verlag, 82 p. + app. 110 p.

3. C. Steinert, 2000Experimentalle UntersuchungenzumAbbrand- und Feuerubersprungs-verhalten von Personenkraftwagen. Vfbd-Zeitschrift, 4941631720042-1804
4. C. Steinert, 2000Feuerubersprung und Abbrandverhalten von PersonenkraftwagenTl. 1. Stuttgart: Fraunhofer IRB Verlag, 95 p. + app. 172 p.
5. CFX User's Guide, available at www.ansys.com
6. E. G. Butcher, G. J. . Langdon-Thomas, G. K. Betford, 1968Fire and Car Park Buildings. Fire Note, 10HMSO, London, 24 p.
7. E. S. Oran, J. P. Boris, 1987Numerical Simulation of Reactive Flow, Elsevier Science Publishing Company, New York, 0-52158-175-3
8. G. P. . Forney, W. F. Moss, 1994Analyzing and Exploiting Numerical Characteristics of Zone Fire Models, Fire Science and Technology, 1449600285-9521
9. I. . Ponce, P. Polednak, 2006Automobile Fires (in Slovak).Proceedings of the 2nd Int.Conf. on Fire Safety and Rescue Services, Zilina, Slovakia, 2692748-08070-539-9
10. I. D. Bennetts, D. J. Proe, R. Lewins, I. R. Thomas, 1985Open-Deck Car Park Fire Tests, BHPMelb. Res. Lab. Rep. No. MRL/PS69/85/001, 24p.
11. J. .Mangs, O. Keski-Rahkonen, 1994Characterization of the Fire Behaviour of a Burning Passenger Car, Part. I: Car Fire Experiments, Fire Safety Journal, 23117350379-7112
12. J. B. Schleich, L. G. Cajot, M. Pierre, M. Brassauer, J. M. Franssen, J. Kruppa, D. Joyeux, L. Twilt, J. . Van Oerle, G. Aurtenetxe, 1999Development of Design Rules for Steel Structures Subjected to Natural Fires in Closed Car Parks. Luxembourg, European Commission
13. J. Ewer, E. Galea, M. Patel, F. Jia, A. Grandison, Z. Wang, S. M. A. R. T. F. I. R. E. -The, Field. Fire, Environment. Modelling, 2010The Fifth European Conference on Computational Fluid Dynamics, ECCOMASCFD 2010 (Pereira, J. C. F., Sequeira, A, Pereira, J. M. C., eds.), Lisbon, Portugal, June 2010, 978-9-89967-781-4
14. J. Kruppa, D. Joyeux, B. Zhao, 1997Evaluation of the Fire Resistance of a Car Park Structure Based on Experimental Evidences, Proceedings of the 2nd Int. Conf. on Fire Research and Engineering (August 1997), Gaithersburg, MD, 416426
15. J. Quintiere, 1984A Perspective on Compartment Fire Growth. Combustion Science and Technology, 3911540010-2202

16. K. H. Digges, R. G. Gann, S. J. Grayson, M. M. Hirschler, R. E. Lyon, D. A. Purser, J. G. Quintiere, R. R. Stephenson, A. Tewarson, 2008Human Survivability in Motor Vehicle Fires. Fire and Materials, 32Iss. 4, 2492581099-1018
17. K. Mc Grattan, B. Klein, S. Hostikka, J. Floyd, 2009Fire Dynamics Simulator (Version 5): User's Guide, NIST Special Publication 10195NIST, Washington
18. K. Okamoto, N. Watenabe, Y. Hagimoto, T. Chigira, R. Masano, H. Miura, S. Ochiai, H. Satoh, Y. Tamura, K. Hayano, Y. Maeda, J. Suzuki, 2009Burning Behaviour of Sedan Passenger Cars. Fire Safety Journal, 443013100379-7112
19. L. Halada, P. Weisenpacher, J. Glasa, 2010Possible Use of Computer Fire Simulation for Automobile Fire Safety Purposes, Proceedings of the 4th Int. Conf. on Fire Safety and Rescue Services, Zilina, Slovakia, 6877978-8-05540-202-8
20. M. . Shipp, M. Spearpoint, 1995Measurements of the Severity of Fires Involving Private Motor Vehicles. Fire and Materials, 1931431511099-1018
21. O. Pettersson, S. E. Magnuson, J. Thor, 1976Fire Engineering Design of Structures, Swedish Institute of Steel Construction, Publication 50
22. P. Polednak, 2010Experimental Verification of Automobile Fires (in Slovak). Proceedings of the 4th Int. Conf. on Fire Safety and Rescue Services, Zilina, Slovakia, 248255978-8-05540-202-8
23. P. Weisenpacher, J. Glasa, L. Halada, 2010Computer Simulation of Fires in Automobile Engine Compartment, Proceedings of the 4th Int. Conf. on Fire Safety and Rescue Services, Zilina, Slovakia, 7887978-8-05540-202-8
24. R. G. Geiwan, 1973Fire Experience and Fire Tests in Automobile Parking Structures. Fire Journal, 6745054
25. R. G. Rehm, H. R. Baum, 1978The Equations of Motion for Thermally Driven, Buoyant Flows, Journal of Research of the NBS, 832973081740-0449
26. S. Galla, 2010Fire Intervention of Fire and Rescue Brigade (in Slovak), Proceedings of the Workshop on Fire Risk and Damage Event Liquidation, Bratislava, Slovakia

27. T. Kitano, O. Sugawa, H. Masuda, T. Ave, H. Uesugi, 2000Large Scale Fire Tests of 4-Story Type Car Park. Part. 1. The behaviour of structural frame exposed to the fire at the deepest part of the first floor. Proceedings of the 4th Asia-Oceania Symposium on Fire Science and Technology (Yamada, T., ed.), May 2000, Tokyo, Japan, 527538
28. V. Babrauskas, R. B. Wiliamson, 1979Post-flashover Compartment Fires:Application of a theoretical model, Fire and materials, 3171099-1018

CITATION

LadislavHalada, Peter Weisenpacher and Jan Glasa (2012). Computer Modelling of Automobile Fires, Advances in Modeling of Fluid Dynamics, Dr. Chaoqun Liu (Ed.), ISBN: 978-953-51-0834-4, InTech, DOI: 10.5772/48600.

Index